CYBERNETICS AND THE CONSTRUCTED ENVIRONMENT

Grounded in contemporary landscape architecture theory and practice, *Cybernetics and the Constructed Environment* blends examples from art, design, and engineering with concepts from cybernetics and posthumanism, offering a transdisciplinary examination of the ramifications of cybernetics on the constructed environment. Cybernetics, or the study of communication and control in animals and machines, has grown increasingly relevant nearly 80 years after its inception. Cyber-physical systems, sensing networks, and spatial computing—algorithms and intelligent machines create endless feedback loops with human and nonhuman actors, co-producing a cybernetic environment. Yet, when an ecosystem is meticulously managed by intelligent machines, can we still call it wild nature? Posthumanism ideas, such as new materialism, actor-network theory, and object-oriented ontology, have become increasingly popular among design disciplines, including landscape architecture, and may have provided transformative frameworks to understand this entangled reality. However, design still entails a sense of intentionality and an urge to control. How do we, then, address the tension between the designer's intentionality and the co-produced reality of more-than-human agents in the cybernetic environment? Is posthumanism enough to develop a framework to think beyond our all-too-human ways of thinking? For researchers, scholars, practitioners, and students in environmental design and engineering disciplines, this book maps out a paradigm of environmentalism and ecological design rooted in non-communication and uncontrollability, and puts a speculative turn on cybernetics.

Zihao Zhang is a designer, educator, and scholar in landscape architecture. He currently serves as an assistant professor in the landscape architecture program at the City College of New York.

CYBERNETICS AND THE CONSTRUCTED ENVIRONMENT

Design Between Nature and Technology

Zihao Zhang

Routledge
Taylor & Francis Group
NEW YORK AND LONDON

Keywords selected from the chapter titles of this book were used as prompts in an AI image generator to create the cover image.

Designed cover image: Created by Shurui Zhang with Midjourney.

First published 2025
by Routledge
605 Third Avenue, New York, NY 10158

and by Routledge
4 Park Square, Milton Park, Abingdon, Oxon, OX14 4RN

Routledge is an imprint of the Taylor & Francis Group, an informa business

Library of Congress Cataloging-in-Publication Data
Names: Zhang, Zihao (Professor of landscape architecture), author.
Title: Cybernetics and the constructed environment: design between nature and technology / Zihao Zhang.
Description: New York, NY: Routledge, 2025. |
Includes bibliographical references. |
Summary: "Grounded in contemporary landscape architecture theory and practice, Cybernetics and the Constructed Environment blends examples from art, design, and engineering with concepts from cybernetics and posthumanism, offering a transdisciplinary examination of the ramifications of cybernetics on the constructed environment."— Provided by publisher.
Identifiers: LCCN 2024007060 (print) | LCCN 2024007061 (ebook) |
ISBN 9781032341743 (hardback) | ISBN 9781032341750 (paperback) |
ISBN 9781003320852 (ebook)
Subjects: LCSH: Landscape architecture—Data processing. | Cybernetics.
Classification: LCC SB475.9.D37 Z43 2025 (print) |
LCC SB475.9.D37 (ebook) | DDC 712.0285—dc23/eng/20240613
LC record available at https://lccn.loc.gov/2024007060
LC ebook record available at https://lccn.loc.gov/2024007061

ISBN: 9781032341743 (hbk)
ISBN: 9781032341750 (pbk)
ISBN: 9781003320852 (ebk)

DOI: 10.4324/9781003320852

Typeset in Sabon
by codeMantra

To Shurui, the first reader of this book.

CONTENTS

FIGURES

Note, all figures are the author's, unless otherwise noted.

ABOUT THE AUTHOR

Zihao Zhang is a designer, educator, and scholar in landscape architecture. Through building transdisciplinary, critical analyses across design, engineering, and environmental humanities, his work interrogates the ramifications of cybernetics on contemporary culture and the constructed environment. As a landscape theorist, he provides critical analyses of the entanglement between nature and technology, the human and nonhuman realms, as well as ecosystems and intelligent machines. His design experiments inspire innovative landscape strategies to address pressing issues such as the climate crisis and social injustice, along with their cumulative effects on communities and urban landscapes. Zihao Zhang currently serves as an assistant professor in the landscape architecture program at the City College of New York.

INTRODUCTION

The Upscaling of the Cybernetic Imagination

We are surrounded by cybernetic machines. They construct the environment we inhabit. We therefore live in a cybernetic environment. This book hopes to investigate the entanglement of humans, nature, and technology in the upscaling of society's cybernetic imagination. What does a cybernetic environment look like, and how do we make sense of it?

DroneSeed, a Seattle-based company, uses drone swarms to plant forests after disturbances such as wildfire. One drone operator is able to propagate 80 acres of trees within eight hours and reforest a post-wildfire site within 30 days. This is six times faster than human-performed planting. The drones' flying paths are pre-programmed and optimized, and they then plant and manage trees semi-autonomously. Robotics has also entered food production, from precision farming to home gardens. FarmBot is a company that combines computer numerical control (CNC) technologies with gardening, producing robotic armatures that automatically monitor plant growth and water, and also prune the plants.

To critique the schizophrenic garden-machine motif within the American psyche, Leo Marx used "the machine in the garden" as a metaphor.[1] Yet, this phrase now takes on a literal sense—real robots working in real gardens. Compared to machines, robots carry with them a sense of autonomy rather than automation. Machines present an image of sophisticated gears and oily chains that automate human physicality; robots are about sensors and computer chips that automate their users' human thinking and produce an appearance of sentience.

RangerBot is an autonomous, vision-based, underwater robot developed to protect Australia's Great Barrier Reef from the prolific, coral-eating, crown-of-thorns starfish (COTS) that threaten the health of the reef

DOI: 10.4324/9781003320852-1

ecosystem. RangerBot is an engineering achievement of machine-learning techniques, robotics, and computer vision. Scientists use a random-forest classifier, a machine-learning algorithm trained on underwater footage, to build a vision-based system in order to detect and track the starfish. The robot then fires a lethal injection and kills the starfish without harming the reef itself.[2] The robot was initially developed as COTSBot by the Queensland University of Technology and the Great Barrier Reef Foundation, with support from Google Impact Challenge. RangerBot extends its predecessors' capabilities and becomes a "Swiss Army knife"-style robotic system that can do more for upscaling environmental monitoring and the management of reef ecosystems worldwide. Aside from pest control, it now comes equipped with further tools and sensors to monitor reef health and map underwater areas at scales previously impossible. From COTSBot to RangerBot, the name change represents a shift of imagination from a straightforward robotic function to a job position that requires independent decision-making. Indeed, numerous human diver-rangers now monitor and manage the reef ecosystem, including manually killing the COTS with injections. In a way, the RangerBot is conceived of as just another diver-ranger. Scientists have attributed increased agency and intelligence to this robot, viewing it as more than an extension of human eyes and arms; it has become an intelligent decision-maker and co-worker in the reef "garden."

From 2017 to early 2021, I joined an interdisciplinary research team of scientists and engineers housed within the Link Lab at the University of Virginia. The team worked with Norfolk, a coastal city in Virginia, investigating cyber-physical systems in coastal climate adaptation.[3] Scientists were advancing techniques on all fronts with a sensing-predicting-controlling feedback loop as an undergirding framework, from sensor networks to machine-learning algorithms and responsive infrastructures. In one project, scientists trained an AI system, using deep reinforcement learning (RL) techniques, to manage a simulation of an urban stormwater system, balancing several competing objectives. This experiment demonstrated that machine learning provides a viable approach for determining policies for real-time control systems. These machine-learning agents could be transferred to cities much like Norfolk, and uploaded into the real-world cyber-physical infrastructure.

Many cities are working with emerging companies that focus on responsive infrastructures. Even though machine learning does not yet replace rule-based control as an industrial standard, research like that in the Link Lab has pushed industry toward a future where machine-learning agents will play a preeminent role in managing urban environments. When the Link Lab's research projects are concretized in one way or another, it will mean that urban environments will be managed not only by humans but also by intelligent agents deeply embedded in the distributed cyber-physical infrastructures. The outcome will be a hybrid of human and machine intelligence, and a meshing of objective functions and goals.

Underpinning these environmental practices is a broad-based cybernetic imagination within contemporary culture. At the heart of cybernetics is the use of self-regulating and recursive causality to analyze control and communicative behaviors within biological and mechanical systems. Cybernetics began as a transdisciplinary field of research in the post-World War II era. Its key principles, such as feedback, self-regulation, and homeostasis, rapidly became cornerstones for modern control theory and systems theory, and have proven useful in designing powerful machines. We witness in present-day environmental practices an upscaling of the cybernetic imagination, in the sense that we desire to either introduce more intelligent machines to manage all sorts of environments or conceptualize an environment itself as a grand cybernetic machine to be controlled and optimized.

Over the past decade, the smart city has become the most ambitious cybernetic project; it mobilizes different social sectors to develop technologies to make cities "smarter." Sensors, platforms, crowd-sourced data, machine-learning algorithms, and artificial intelligence—a plethora of new tools are being developed to experiment with feedback loops between urban systems. The goal is to make cities "smarter," but what counts as smart? The word is often narrowly defined as more efficient and connected, making urban processes faster and growing larger urban datasets. Though the notion of a smart city has been extensively criticized from the beginning, such criticism has not arrested the urge to optimize cities as cybernetic machines.

In February 2024, as I was finishing this book, Apple ushered in the "era of spatial computing" with the introduction of Vision Pro, a pair of goggles that transcends traditional virtual and augmented reality technologies. Its sensors, 3D cameras, and processors capture, reconstruct, and live-stream the surrounding environment—or any imagined environment—and project user interfaces like web browsers and emails into this space. Apple defined it as a technology for spatial computing because the surroundings appear computable and controllable through simple pinches. While smart cities turn sensors inside out to make computation the background of the environment, spatial computing folds the environment outside in to immerse its users in a truly "smart environment." The introductory video showcases a scenario where a user engages in mindfulness practices within a digital world replete with abstract geometries and vibrant colors. In a world already saturated with information, a paradoxical solution for tranquility is to inundate ourselves with even more information to craft a hyper-reality. In this constructed environment, you decide whether something should be projected in front of your eyes; you are in ultimate control.

"A city is not a computer" and "urban intelligence is more than information processing."[4] Many critics, like Shannon Mattern, have voiced clear objections. However, these claims will not end cybernetic thinking's contagious nature; as long as one can conceptualize a system with measurable input and output, then an entire branch of mathematics and multiple techniques in modern control theory are available to make this system appear to be computable, controllable, and optimizable. This is partly why philosopher Yuk Hui notes that cybernetics "remains a thinking of totalization, since it aims to absorb the other into itself."[5]

Cybernetics Unexplored?

Within multiple cybernetic practices, we begin to see abnormalities that fail to fit within the mainstream cybernetic imagination that is characterized by increased communication and improved control. For example, in the Link Lab's research, one RL agent developed its own understanding of an urban drainage system without human knowledge and devised certain unexpected strategies beyond scientists' comprehension. "Can machines think?" Alan Turing inquired when computers were in their infancy.[6] Now, we must pose a second question: "Can machines think what humans cannot?"

Relying on machine-learning techniques such as RL, scientists have trained intelligent agents that not only outperform humans in various aspects but also develop their own strategies with a "machine flavor." For example, AlphaGo, developed by Google's DeepMind, is an RL agent specializing in the strategy board game Go. This game involves two players who each try to surround more territories than their opponent. Over the past years, AlphaGo, with its predecessors Zero and Master, has vanquished many of the best human Go players. Most importantly, it has developed unique strategies that professional Go players have never before seen.[7] The game of Go was invented approximately 2,000 years ago, and humans have explored the game long enough to believe that all strategies had been identified. Yet AlphaGo has expanded the game's possibilities with a machine perspective.

Similar examples can be found in the field of video games. AlphaStar is another RL-based agent that has reached the grandmaster level (the highest rank reachable by competing with other players) in the real-time strategy video game StarCraft.[8] The AI community regards this experiment as a breakthrough, because real-time strategy games such as StarCraft are infamous for their vast action space, containing a planning horizon of thousands of real-time decisions with imperfect information. AlphaStar not only mastered the game but also developed a unique playing style. Consider this comment made after one professional human player competed against AlphaStar:

> AlphaStar is an intriguing and unorthodox player—one with the reflexes and speed of the best pros but strategies and a style that are entirely its own. The way AlphaStar was trained, with agents competing against each other in a league, has resulted in gameplay that's unimaginably unusual; it really makes you question how much of StarCraft's diverse possibilities pro players have really explored.[9]

Here, we should overcome the human-machine rivalry trope popularized by pop culture, and instead focus on how humans and machines explore problems and events such as games together, and how emergent behaviors arise in the process of interaction. In both Go and video games, these behaviors are embodied in new ways to play old games.

There are examples of less "intelligent" machines. For example, Sougwen Chung is an artist known for her drawing series, where she collaborates with robots and computer-vision algorithms. Her set-up was simple: a table with a large piece of canvas, the artist on one side and a robotic arm on the other. They then performed a drawing duet. With computer vision, the robotic system traced the artist's movements and attempted to reproduce her gestures with the robotic arm, holding a pen. Yet due to the machines' imperfections—which caused jittering and delay—the artist was forced to adjust her movements in response to the machine. The feedback loops and recursive processes between the machine and the artist did not result in homeostatic and expected results, but rather in drawing styles which the artist admitted she could not have produced on her own.

Moreover, the past several years have seen a range of prototyping research in landscape design, which relies on cybernetic technologies and recursive principles but produces results beyond cybernetic control. For example, landscape architects have used physical hydro-sediment tables combined with sensor arrays and actuators to explore land-building strategies. By manipulating hydrological patterns in real-time, and utilizing sensing-processing-actuating feedback loops, the machine assembly constructs new landforms beyond the designers' intent and control.[10] These cybernetic machines open a liminal space of exploration between designers' intent and a co-produced reality resulting from the interactions of many agents.

There are further instances where cybernetic machines make us question how much we have explored with our limited, all-too-human ways of knowing and thinking. With these examples, one cannot help asking: are these practices still cybernetics? If repetitive actions and recursive causality do not lead to communication and control, how do we define the results of these cybernetic mechanisms? In addition, if cybernetic machines could help expand our understanding of the environment, what possibilities lie in the cybernetic environment? Jennifer Gabrys, the author of *Program Earth*, envisions "planetary computerization" as providing opportunities for speculative experiments, for "propositioning, instigating, and triggering—beyond the usual automated sensor-actuator triggers of cybernetics—toward indeterminacy and openness."[11] The upscaling of society's cybernetic imagination gives rise to competing versions of cybernetic environments. We should cultivate those that help us reimagine the environment as a reserve for possibilities and a future we cannot now foresee.

Three Tensions within the Cybernetic Environment

Before these cases become truly transformative, we face an acute problem: many of these cases lie outside our interpretive and evaluative categories based on optimization and control. This challenge may be described as three perceived tensions specific to the cybernetic environment.

First, the cybernetic environment presents an upscaling entanglement between nature and technology, biological and technological, machine and wildness, biotic and abiotic. In the RangerBot case, can we still call the Great Barrier Reef an area of wild nature when it is carefully managed by humans and intelligent machines? Should we instead refer to it as a garden, a more apt term for cultivated nature? Are concepts such as *hybridity*, *cyborg*, and *system coupling* sufficient to describe an entangled reality?

Second, in many of the above examples, the machines are more than tools that simply automate human labor; they exhibit different levels of autonomy, as seen in the Link Lab's RL agent. Thus, a challenge arises: do we trust these robots—which occasionally act in unexpected fashions—enough to cede the environment to them? The environment is not a game of Go or a video game with a reset button. Deep in our imagination lurks the scenario of AI gone rogue. Underlying this discomfort is a deeper predicament for discussing nonhuman agency. Contemporary environmental humanities and political ecologies tend to attribute agency to more-than-human objects—machines, animals, plants, organizations, and policies. Intellectuals want to show a sense of humility as human beings, recognizing other forms of agencies that co-produce the environment. However, the irony is that the "agency" we attribute to others is an observed efficacy. After all, the so-called nonhuman agency is effectiveness; these nonhuman beings can be used to benefit human goals, reinforcing another level of human hubris and control. Therefore, the tension lies between the willingness to embrace a sense of humility, and the inability to develop a framework to think beyond our all-too-human ways of thinking.

Third, the cybernetic environment highlights a space between intent and reality, posing threats to the act of design itself. Many contrast the sciences with design—the former is about revealing patterns in the world, while the latter is about rearranging things and relationships to create new patterns. Design entails a sense of intentionality and an urge to control how things ought to be and become; it achieves goals and constructs an ideal end. However, within the cybernetic environment, humans are not the source of agency, and our actions are modified and interfered with by other goals. In Chung's drawing practice, she as the artist must constantly adapt to the machine's movement. The feedback loops between human and nonhuman agents will lead to a direction outside anyone's plan. How do we then conceptualize the act of design itself, which embodies an ultimate form of idealistic thinking and human control? Thus, the third tension is the one between the designer's intentionality and the co-produced reality of more-than-human agents in the cybernetic environment.

This book addresses these tensions through two conceptual moves. First, I will deploy cybernetics as a lens through which to group a range of extremely diverse practices. Then, I will envision a field called "cybernetic environment," in which machines and cybernetic thinking play a preeminent role in constructing the environment. Second, I will analyze these practices, and, therefore, their underlying cybernetic thinking, in light of contemporary posthumanist arguments on nonhuman agency and various ontological concerns. In a way, this book extends philosopher Yuk Hui's claim that "we are in an epoch of cybernetics" by asking what exactly this "cybernetic epoch" looks like when humans are removed from the source of agency, and when the environment is understood as a co-produced result involving nonhuman agents. It is about exploring a version of posthumanist cybernetic thinking. This alternative view of cybernetics transforms how we understand our relationship with machines and other nonhuman species in a shared environment and a co-produced future.

Before we proceed, we shall focus on each conceptual move separately—cybernetics and posthumanism.

Why Cybernetics? From Cybernetics to Posthumanism

We must recognize that there exists a deep misconception about the term *cybernetics*. The term is often used interchangeably with *robotics*, and occasionally misunderstood as modern control theory itself. It is often associated with linear mechanisms of stability and control, and thus is often critiqued by those who embrace nonlinear thinking and indeterminacy. Even Jennifer Gabrys appears to equate cybernetics to determinacy and closeness, as stated above. These misconceptions and critiques of cybernetics are understandable, because cybernetic principles underpin all modern machines and control systems. However, the history of cybernetics reveals that cybernetic thinking is extremely diverse. The above examples indicate that certain aspects of cybernetics remain underexplored.

Many agree that cybernetics began with a series of interdisciplinary meetings held in New York from 1944 to 1953, known as the Macy Conferences on Cybernetics; they brought together important postwar intellectuals, including mathematicians Norbert Wiener, Claude Shannon, and John von Neumann; anthropologists Gregory Bateson and Margaret Mead; neurophysiologist Warren McCulloch; physicist and philosopher Heinz von Foerster; and psychiatrist W. Ross Ashby. As Wiener's book title—*Cybernetics: or Control and Communication in the Animal and the Machine*—suggests, the field of cybernetics is transdisciplinary. Wiener aimed to study recursive and self-regulatory mechanisms in biological systems and machines. To some extent, at the core of cybernetics is the study of a simple function: $y = f(y)$; the output of the system feeds back to itself as input. This period of the cybernetics movement is also known as first-order cybernetics, to distinguish it from the later, second-order cybernetics that began in the late 1960s.

Without a doubt, many early cybernetic principles, including feedback, recursive causality, and self-regulation, became concepts that underpinned modern disciplines such as computer science, control engineering, and artificial intelligence.[12] Even the most cutting-edge AI research, such as deep RL, a machine-learning framework that successfully trained AI systems such as AlphaGo, is, at base, cybernetic. The RL framework conceptualizes a machine agent that observes its environment, evaluates its state, devises strategies to change the environment, and then repeats the entire process. The output of the RL agent is a modified environment, which then feeds back to the agent as new input for its next action. Recursive causality is an elegant conceptualization for building powerful machines; all modern machines are cybernetic machines.

We must recognize that early cyberneticians viewed cybernetic machines differently. Cybernetic artifacts, including William Grey Walter's cybernetic tortoises, Wiener's "moth and bedbug," and Von Neumann's cellular automaton, were constructed during the Macy Conference era. Yet, these machines were merely demonstrations of cybernetic thinking; they were physical diagrams of different types of feedback systems. The cyberneticians' goal was not to build powerful machines, but to construct a universal framework that could cut across materially and physically different entities. From this perspective, cybernetics has been narrowly interpreted as a theory for building powerful machines for control.

Before the Macy Conference era, Norbert Wiener developed the cybernetic thinking that would be used during World War II. His goal was to build a machine—the "antiaircraft predictor"—to predict an enemy pilot's flight path and launch an antiaircraft shell to down the plane. However, as pointed out by historian Peter Galison, Wiener quickly realized the more profound implications and universality of his conceptual framework based on recursive causality, and therefore wanted to direct cybernetics toward a philosophical concept.[13] Moreover, with cybernetics' history in warfare, Wiener's conscience led him to steer cybernetics toward a theoretical endeavor rather than the practicalities of building potentially destructive machines.[14]

In charting this history, we can also recognize that communication and control have been two key concepts since Wiener's coinage of the term *cybernetics* (in Greek, "steering person" or "helmsman") in the summer of 1947, to denote what he hoped would become "a new science of control mechanisms in which the exchange of information [or communication] would play a central role."[15] In other words, before cybernetics, engineers would decipher and design the internal structure of a system in order to build control mechanisms. In contrast, cybernetic thinking constructs a black box, ignoring the system's internal structure and focusing on the system's input and output information as the core to achieving a sense of control; communication mechanisms become the fundament for control strategies. This conceptualization became the foundation of modern control theory, based on measuring input and output in time sequences, and computing control strategies. From this vantage, this book pulls together a range of cybernetic practices that lie outside the core conceptualization of cybernetics based on control and communication. We will see non-communicative behaviors and uncontrollable situations in some of the examples presented here. They suggest that there may be much unexplored in cybernetic thinking itself—a version of cybernetic thinking that is non-communicative and less about control.

Due to its interdisciplinary nature, cybernetics has been imported into and interpreted within many fields of research in the twentieth century, including landscape architecture. Ian McHarg imported cybernetic thinking into landscape architecture via ecological science. In his seminal work, *Design with Nature*, McHarg discussed "entropy" in a way that echoes Wiener's interpretation; McHarg viewed entropy as destructive to homeostasis, stability, and ecological fitness.[16] However, "misinterpretations" occurred as well. When introducing his "RSVP scoring design framework" in the early 1970s, landscape architect Lawrence Halprin wrote, "scores communicate but do not control."[17] He also argued that "one of the gravest dangers that we experience is the danger of becoming goal-oriented."[18] The clear reference to Wiener's book title and the critique of means-end reasoning suggest that Halprin found the recursive thinking of cybernetics useful, but wanted to interpret it differently, from a landscape sensibility.[19]

Halprin made these provocative claims in the 1970s because, on the one hand, like other landscape architects, he worked with uncontrollable living materials; on the other hand, unlike McHarg, who focused on large-scale landscape planning through a scientific lens, Halprin needed to work with people throughout the design processes—or, from a present-day perspective, conduct community-engaged design.

Indeed, in cybernetic thinking from its early stages, humans were regarded as a predicament. The roles assumed by observers and engineers in conceptualizing the cybernetic model became a key issue explored during the late Macy Conference era, but without success.[20] Early cyberneticians, such as Bateson, noted that the issue of reflexivity is more than subjectivity in modeling, but is, instead, an issue in the epistemological realism that undergirds early cybernetic thinking, and "the problems posed by including the observer could be addressed only if a substantial reworking of realist epistemology was undertaken."[21] Ultimately, a sense of liberal humanism supported the unwillingness to demote humans from their privileged position in conceptualizing consciousness, agency, and intelligence. Cyberneticians' question became how to arrive at a theory that included observers' roles in the cybernetic systems they observed.

Thus, the field of cybernetics began to diverge. Practice-oriented scholars, such as engineers and computer scientists, shifted their attention from theoretical inquiries and toward developing powerful machines with cybernetic principles. Many fields, such as systems engineering, computer science and artificial intelligence, grew into independent disciplines with specific questions and concerns.[22] In the meantime, those who identified themselves as cyberneticians wanted to distinguish their work from the practice-oriented disciplines, and decided to question the unresolved observer issue, attempting to develop a new cybernetic framework. In 1967, anthropologist Margaret Mead, a key participant in the Macy Conferences, addressed the American Society for Cybernetics about the need for a recursive application of cybernetics on itself, treating the observer as one cybernetic system constructing models of yet another cybernetic system.

The significance of this conceptual move is more than a new model of cybernetics; rather, it is a path toward a series of reflections on human agency and control. This is why many trace the cybernetics movement as a major tenet contributing to the twenty-first-century posthumanism movement across disciplines.[23] In her seminal work, *How We Became Posthuman*, literary critic N. Katherine Hayles schematized this genealogy into three waves of cybernetics that traveled between different fields of study. Each wave concerns different aspects of the feedback mechanism. The first wave denotes the Macy Conferences era, with its focus on self-regulating and homeostatic systems via feedback mechanisms, or $y = f(y)$. Hayles attributes the second wave of research to the second-order cybernetics movement, which began in the late 1960s. Its major development was to conceptualize the observer as an observing system. This is why Heinz von Foerster also called it "cybernetics of cybernetics"—a recursive application of cybernetics on itself.[24] Second-order cybernetics marked a reflexive turn in epistemology, from epistemological realism to constructivism.[25]

One of the most important concepts of the second wave of research was the autopoiesis theory developed by Chilean biologists Humberto Maturana and Francisco Varela.[26] By studying frogs' eyes and nervous systems, the biologists found that the fibers between the eye and the brain pre-process information so that the brain does not receive an entire package of image data from the eye; a frog's nervous system constructs a reality for its brain to ensure the frog's survival.[27] Maturana and Varela asserted that autopoietic systems, including those of humans, use inputs to reconstruct their own system organizations, thus maintaining system identities. "The living organization is a circular organization which secures the production or maintenance of the components that specify it in such a manner that the product of their functioning is the very same organization that produces them."[28]

The autopoiesis theory presents a completely distinct conceptualization of agency, intelligence, and consciousness. Within the autopoiesis framework, these human concepts become epiphenomena which we produce with our system operations, in order to maintain the organizations of a class of special autopoietic systems we call *Homo sapiens*.

Hayles extended cybernetics into the 1990s by attributing the third wave of research to the field of artificial life that focused on the notion of emergence.[29] To some extent, cybernetics has always been explored in a paradigm of homeostasis, in which stability and equilibrium have been accepted as default goals for feedback mechanisms. This can be seen from Wiener's conceptualizations of entropy, equilibrium, and meaning in *The Human Use of Human Beings*, originally published in 1950, in which Wiener tried to push cybernetics into understanding human society. "In control and communication, we are always fighting nature's tendency to degrade the organized and to destroy the meaningful; the tendency…for entropy to increase."[30] Evidently, Wiener associated a sense of moral imperative in maintaining a system's stability. Indeed, in building powerful machines and control systems, defining a stable state is a priori for developing control strategies that stabilize the system. Autopoiesis reflects observers, yet still in a homeostatic paradigm, asserting that observers are cybernetic systems struggling against disorganization.

Hayles's third wave of research reflected the notion of homeostasis and stability, and investigated spin-off behaviors. In the 1990s, multi-agent-based simulations suggested that complex behaviors can be achieved by recursively exercising simple rules. Humans, to a certain extent, become "computer programs" in a "computational universe."[31]

Hayles's observation was acute. However, we must read her interpretations historically and critically. Her presentation of the development from early cybernetics to posthumanism is easily misinterpreted as paradigm shifts, with one understanding replacing the previous one; early cybernetics was replaced by autopoiesis (from epistemological realism to constructivism), and homeostasis was replaced by emergence (from stability to spin-off behaviors). We must recognize that this is a specific genealogical construction of cybernetics in the service of Hayles's argument for a path toward posthumanism. Its history involved more of a thickening process, with multiple interpretations absorbed into the umbrella term of "cybernetics" and its recursive framework. Thus, many would today use the term cybernetics to denote a collection of ideas, without distinguishing between first- and second-order cybernetics.[32]

We may frame the development of cybernetics differently. At the heart of cybernetics is recursive causality—the output of a system decides the input. However, how we interpret this feedback mechanism depends on the episteme in which cybernetic thinking has been explored, in both theory and practice. For example, the second wave of cybernetics that posed questions concerning the observers' role mirrored a broad-based reflection on authorship, objectivity, and scientific truth in the mid-twentieth century, across the sciences, arts, and humanities. In science history, Thomas Kuhn's 1962 book *The Structure of Scientific Revolutions* found its popularity beyond academics, as did the term "paradigm shift." Kuhn presented an alternative view of the "great man" history of scientific and technological revolutions, arguing that scientific truth was constructed by the scientific community based on a consensus with regard to the underlying assumptions, techniques, and values shared by that community.[33] The intellectual excellence of individual scientists was questioned; scientists themselves, with their unconscious minds, became the subjects of study.

Similarly, problematized authorship was a key issue in early twentieth-century works of art. Marcel Duchamp's readymade sculpture "Fountain" (1917) was one of the earliest challenges to artists' roles in the production of art. Conceptual artist Sol LeWitt's wall drawing series essentially consists of written instructions for drawings on walls, instead of the drawings themselves.

Similar concerns may be found in the humanities, as well. In his 1967 essay "The Death of the Author," Roland Barthes questioned the role of authors in literature and argued that an author was not a creator of meaning in the text, but was only one who combined different texts.[34] In a similar vein, Michel Foucault regards authorship as an "author function," which speaks to the need to pin a discourse to a person, as a regulator of meaning. This reflection can be situated within Foucault's thinking, encapsulated in his seminal work *The Order of Things*. Toward the end of his book, Foucault provocatively claimed that the appearance of a human figure

> was the effect of a change in the fundamental arrangements of knowledge. As the archaeology of our thought easily shows, [hu]man is an invention of recent date. And one perhaps nearing its end … like a face drawn in sand at the edge of the sea.[35]

With these examples, we should recognize that second-order cybernetics and autopoiesis theory form a single stream within a broad-based reflection on human exceptionalism across fields, and the emergence of posthumanism toward the late twentieth century. One unique aspect of cybernetics is that it has provided a systemic account for the reflection of human agency, with special attention to the issues of control and communication.

Similarly, the notion of emergence is not specific to the field of cybernetics. Toward the late twentieth century, many fields, including landscape architecture, started to explore the notion of emergence. Since the late 1990s, landscape architects have focused on the ecological emergence and becoming of landscapes, through theory and practice, framing landscapes as processes of unfolding and leading to new forms of relationships between biotic and abiotic beings in the environment.[36]

On another level, Hayles's work demonstrates that cybernetics, as transdisciplinary thinking by nature, navigates between different fields of research. Because of its transdisciplinary nature, Hayles can extend cybernetics to the artificial life research of the 1990s, exploring cybernetic thinking with the notion of emergence at its early stage. Nearly 30 years have passed since *How We Became Posthuman*. Today, we are better positioned to consider notions such as emergence and open-endedness, with transformative cases across cybernetic practices. Moreover, since the 1990s, posthumanism has made its way into intellectual life and has provided a completely new backdrop for considering cybernetic thinking and practices.

Posthumanism

The second conceptual move situates cybernetics within a posthumanist framework, and explores what cybernetics means, with contemporary concerns about nonhuman agency and intelligence. Yet, what is posthumanism?

Cary Wolfe attributes posthumanism to at least two genealogies, one rooted in early cybernetics and the other in Foucault's pronouncements on the appearance and disappearance of "human" as arrangement and rearrangement of knowledge.[37] However, as we have seen, reflections on human agency and exceptionalism were shared across arts, sciences, and humanities in the second half of the twentieth century. Thus, the downside of tracing genealogies is in overlooking the similarities of ideas across fields, and the establishment of an idea as a transdisciplinary effort.

Posthumanism in this book takes on its most general sense and speaks to extremely diverse fields of thinking, including earlier reflections on human agency, and contemporary reflections, such as object-oriented ontology (OOO) and speculative realism. Though certain OOO proponents would distance themselves from posthumanism, their arguments are based on a narrow definition. If we focus on the undertaking of these different ideas, we see that they share a common goal: to challenge anthropocentrism by removing humans from the center of the source of agency. From this perspective, as long as an idea shares a similar sentiment, this book would, without hesitation, categorize it as a version of posthumanist thought. Under this large conceptual umbrella, we may focus on their different approaches for demoting humans from a privileged position for interpretation.

From this perspective, this book offers two alternative perspectives on posthumanism specific to the cybernetic environment: first, considering the entanglement between the human and nonhuman realms, and second, contemplating the complex relations between humans and technology. Tracing the evolution of conceptions of nature and technology through the twentieth and early twenty-first centuries, this book presents a growing reflexivity regarding human agency against nonhuman species and machines, recognizing other forms of agency and intelligence in co-producing the environment.

Moreover, this research seeks to reframe the relationship between cybernetics and posthumanism. Hayles mapped a path from cybernetics to posthumanism in 1999, but after some 30 years of exploration, posthumanism itself has developed from its early stage of challenging human agency to a "nonhuman turn" that focuses on nonhuman agency, and is now moving toward a "speculative turn" that seeks to surpass agency—the human-centered concept itself. In other words, Hayles's work has contributed to a broad-based movement to completely reframe cybernetics itself. At such a level, this book is about situating cybernetics, a 70-year-old concept, within contemporary concerns on nonhuman agency and speculative ontology, and provides a new interpretation that acts in the service of openness and indeterminacy, instead of controlled stability via optimized communication.

The term "nonhuman turn" entails a response to the initial reflection on human exceptionalism. In the late 1990s, scholars in social sciences and the humanities began to turn to the nonhuman realm as an object of inquiry, to challenge humans' position in conceptualizing agency. Narratives about nonhuman agency can be found in science and technologies studies (STS), feminist materialism, animal studies, environmental humanities, and political ecology, among other disciplines. Ideas such as actor-network theory (ANT), companion species, systems theory, and assemblage thinking are influential tenets that have provided analytical frameworks in contemporary posthumanist works.

However, these explorations quickly ran into a paradox that undercut the initial moral incentives. These "nonhuman frameworks" turn nonhuman objects into actants in narrators' stories; in the end, "nonhuman agency" becomes observed efficacy or effectiveness based on human standards and in the service of human agency, while intrinsic value as a nonhuman being in itself is still largely outside human-centric frameworks. The search for "nonhuman agency" ironically leads to another level of human hubris. To use OOO proponent Graham Harman's words, these types of reflection "remain human-centered no matter how many material entities they summon in the night to mould and shape human beings."[38]

The reflection on "nonhuman agency" leads to present-day philosophical concerns under the banner of speculative realism. This "speculative turn" is a reflection on "the prevalent tendency with Kantian and post-Kantian thought to treat the relation between thought and world as the primary subject matter of philosophy."[39] In Harman's terms, it can be understood as a rejection of "access philosophy," specifically the philosophy of *human* access.[40] The epistemic undergirding of nonhuman frameworks, such as ANT and assemblage thinking, still privileges the capacity of (human) knowing as fundamental for philosophical projects, even though they desire to subscribe to a posthumanist mode of thinking. In the end, the core of speculative realism is a critical stance toward contemporary posthumanist projects in terms of their "inability—or better, unwillingness—to create a speculative ontology which moves beyond the narrow confines of what is given to our all-too-human modes of understanding."[41]

In a way, the contemporary posthumanist movement is deeply rooted in early cybernetics, but how do these new realizations, in turn, reframe cybernetic thinking? How can we interpret emerging and unorthodox cybernetic practices differently, with new concepts and vocabularies? Finally, how does the discourse of design, which is essentially a form of speculation, contribute to this debate? These are the questions which this book seeks to address.

The Nonhuman Turn and Landscape Design

Beyond the two conceptual moves, contemporary landscape theory and practice serve as a backdrop for this research. As a profession building its mode of practice by drawing ideas and concepts from neighboring and outside fields, landscape architecture has absorbed diverse arguments and concerns, including cybernetics and posthumanism. As noted above, since the 1960s, cybernetic thinking has been imported into landscape architecture through multiple venues, including ecological science and art. Landscape architects have interpreted major concerns such as homeostasis, feedback loops, and emergence through theory-making and material practices.

One of the goals of this research is to bring landscape theory and practice into today's intellectual life, and interpret the meaning of landscape design in contemporary culture. There are two motivations for this undertaking. On the one hand, *landscape* (as both noun and verb) is a cultural product as well as a cultural technique that has long been an object of inquiry across disciplines. Historians have explored and interrogated society's narratives of nature and technology by using landscape as a lens to unpack the dynamic relationships between these terms.[42] French philosopher François Jullien also turned to "landscape" as a motif to compare Western and Chinese thought on philosophical concerns.[43] However, this type of research fails to recognize that modern landscape architecture, as a young discipline, has developed a body of scholarship through texts and material practices that may provide transformative vocabularies and frames to contribute to intellectual concerns.

On the other hand, since the late 1990s, landscape as a mode of thinking and practice has been turned into a model and adopted by many neighboring design disciplines, including urban design and architecture. Landscape design has been "rebranded" into different doctrines, such as "landscape urbanism" and "ecological urbanism" in contemporary design discourse.[44] Over the past two decades, many landscape architects and theorists have surpassed earlier concerns, such as "process" and "emergent ecologies," and asked more difficult questions about open-endedness and the liminal space between designers' intentionality and co-produced reality. When Bradley Cantrell and Justine Holzman conceptualized the role of responsive technologies in landscape design, they also revealed a larger philosophical project for contemporary landscape architects, writing,

> While the argument for ecological, non-deterministic strategies seemingly holds the answers to the problems of the contemporary landscape, it then becomes problematic to find methods that actually resist deterministic outcomes. At what point are goals, scaffolds, and protocols actually open-ended?[45]

The past several years have seen a nonhuman turn in landscape theory and practice that mirrors the broad-based posthumanism movement. Many posthumanist scholars, including Donna Haraway, Bruno Latour, Graham Harman, and Timothy Morton, have been widely read within the landscape discipline. Landscape designers and theorists are exploring various posthumanist frameworks as ways to articulate their practices. Animals, plants, infrastructure, and land use policies are all treated as active agents; landscape architects choreograph them to envision the unfolding and becoming of the landscape dynamics. To use landscape architect Brian Davis's terms,

> if a space is a landscape, then all of its objects and their dynamic relations are instruments [or agents], but not dumb drills, retaining walls, and land use policies. Rather, they are dynamic objects in relation to one another within a bounded territory containing some measure of human intent.[46]

As a profession and discipline that works with living materials and processes that may extend beyond designers' control, landscape architects have always explored a mode of practice and a type of language that navigates between a desire to control and an uncontrollable reality constructed by more-than-human agents—living and nonliving, material and immaterial. Many contemporary landscape ideas have become inspirational. For example, as a notion gaining currency among designers, *choreography* may help conceptualize the relationship between designers and other agents, including intelligent machines, in a co-produced environment. Interpreting contemporary landscape projects in conjunction with other cybernetic practices within a posthumanist framework may articulate a different version of cybernetic thinking, one that operates beyond "controlled stability" toward a "cultivated wildness."

A Critique of Dualism

To an extent, and on many levels, this book is a critique of binary or dualistic thinking. First, it teases out the profound paradox whereby we view nature and technology as conflicting categories. Indeed, despite critiques of the culture-nature binary thinking issued by many postmodern scholars, contemporary environmental practices still largely rely on oppositional terms to describe intrinsically complex strategies: green/gray; hard/soft; nature-based/artificial; landscape/engineering; ecological/technological; biological/mechanical; human/nonhuman; and living/non-living. The problem of perceived boundaries between technology and nature results in unnecessary and occasionally problematic rationales that prevent us from seeing the environment as a mesh of different frameworks—some human, others not. Most importantly, the dualism between the ecological and the technological allows us to view machines only as layers of infrastructure with which humans extend control over the nonhuman realm, overlooking a version of the environment laden with machine intelligence waiting to be mobilized by designers toward further indeterminacy and openness.

Many ideas, including cybernetics, appear to overcome this dualism. Cybernetic thinking bypasses the perceived division between the mechanical and the biological because cybernetics is organicism rooted in reflections on the mechanism in science.[47] It relies on feedback and information as key concepts used to analyze the self-regulating behaviors of all beings, living or nonliving, biological or technological. As a descendent of early cybernetics, systems theory provides an even more universal concept that can be used to understand all beings by understanding their abstracted components and relationships and ignoring their material and physical forms. Aside from systems thinking, other popular ideas such as cyborg, assemblage, and actor-network all provide frameworks that may overcome the perceived dualism between biology and technology, and nature and technology. However, have we genuinely overcome dualist thinking itself?

As Hui pointed out, cybernetics, among many others, are concepts of totalization, viewing differences as a motivation toward synthesized identity.[48] By constructing a universal view such as cybernetics, systems, assemblages, objects, we construct an antithetical pair on the other end of the conceptual field that is non-cybernetic, non-systemic, and non-object. We are thus still entrapped in the dualist thinking embodied in the Hegelian logic of thesis, antithesis and synthesis.[49]

Therefore, the second aspect of critiquing dualism is to reflect on our habitual mode of thinking, which seeks to find universal explanations. It is also about learning to embrace and live with a multifaceted framework. In a way, this book reflects and critiques cybernetics, not by constructing an alternative universal frame to replace mainstream cybernetic thinking, but by providing a way to interpret cybernetic thinking in different lights, to reveal its multifaceted potential and contradicting interpretations.

Chapters

This book is organized in three parts. Part I, The New Machine in the New Garden, investigates the entanglement of nature and technology by enquiring how they became a pair of antithetical ideas in our conceptual frameworks for understanding the environment. In Chapter 1, I argue that a certain formula underpins today's mainstream environmental narratives: humans use technology to transform nature into habitable landscapes. This transformation formula justifies human agency in altering the environment, and technology becomes an effective means to achieve this end, amplifying and augmenting human capacities in communication and control. To deconstruct this formula, I propose a human-technology-nature triad as a conceptual frame to investigate ideas about three pairings: nature/technology (Chapter 2), human/technology (Chapter 3), and human/nature (Chapter 4).

Chapter 2 traces the nature-technology edge of the triad. Building on historians David Nye and Leo Marx, I argue that three grand themes exist to resolve the perceived conflict between nature and technology in the North American context: (1) pastoralism and garden-machine motif, (2) wilderness and technological sublime, and (3) Anthropocene and ecomodernism. Chapter 3 ventures onto the human-technology edge and investigates how technology historians, science technology, and societyscholars, and technology philosophers conceptualize different models to frame the relationship between human and technology. In a similar manner, Chapter 4 presents different perspectives on the relationship between human and nature as three waves of reflection, with various conceptual models being developed in different fields to challenge, critique, or complete their received models.

Part I concludes with Chapter 5 and its discussion of the convergence of ideas about nature and technology in the twenty-first century. Both discourses have adopted some versions of posthumanist assemblage thinking. This can be described as a coproduction thesis; humans and nonhumans have always co-produced with each other, and humans have always co-evolved within a network of human and nonhuman assemblages and systems, including animals, plants, machines, languages, and cultures.

How can we understand this process of coproduction? Part II, Posthumanism, the Environment and Intelligent Machines, dives into contemporary posthumanism ideas and uses them to unpack the coproduction thesis from a theoretical perspective. Chapter 6 will review three schools of posthuman thought—ANT, assemblage thinking, and OOO—to map the contemporary theoretical terrain in understanding interactions between human and nonhuman species in the shared environment.

Chapter 7 argues that searching for nonhuman agency led to a speculative turn rooted in a critique of "philosophies of (human) access." Ultimately, posthumanism could be described as a project "to create a speculative ontology which moves beyond the narrow confines of what is given to our all-too-human modes of understanding."[50] Then, using recent landscape design research findings as examples, I will argue that "design" is an approach to speculative ontology.

How does the notion of machine intelligence shift when we take on a co-productive framework in understanding the cybernetic environment? Chapter 8 explores contemporary posthuman cognition on agency and intelligence, providing a theory-rich reflection on how we could conceive of intelligent machines such as AI systems in the shared environment. Recent artificial intelligence breakthroughs, such as AlphaGO and ChatGPT, provide many transformative cases that challenge the notion of human-machine intelligence. Through examples from art and design research, this chapter presents a framing of the co-produced environment in which intelligent machines are active co-creators rather than mechanisms to extend human agency and control over the environment.

Part III is a reflection on cybernetics in light of posthumanism and the co-productive thesis explored in previous chapters. Chapter 9 introduces the history of cybernetics in parallel with the development of contemporary landscape theory and practice since the mid-twentieth century. This chapter is an attempt to trace the evolution of ideas across disciplines, with two goals. First, it introduces cybernetics and related concepts to readers unfamiliar with the topics. Second, cybernetics serves as an anchor that connects contemporary landscape architecture discipline to a broader intellectual discourse, and mitigates the disciplinary gaps.

Chapter 10 investigates cybernetics through a mainstream sensing-modeling-actuating framework, issuing critiques of this model by reframing the cybernetic process with another set of terms—coding, choreographing, and attuning. The following chapters use cases in art, design and engineering to replace "sensing" with "coding" (Chapter 11), "modeling" with "choreographing" (Chapter 12), and "actuating" with "attuning" (Chapter 13). This reframing places a negative turn on communication and control. In a way, cybernetics has always been interpreted within a framework rooted in philosophy of (human) access, which reasons that things are communicative, and communication inevitably leads to control. This book tries to reinterpret cybernetics with a speculative ontology, which posits that things are inaccessible and thus non-communicative and uncontrollable in the first place.

With this new conviction, Chapter 14 concludes Part III by proposing two concepts: cultivated wildness and speculative ecology. "Cultivated wildness" reconfigures the notion of control and agency. It highlights the verity in the cybernetic environment that when we seek to control what is considered wild, we are, at the same time, constructing it. Since things are inaccessible and withdrawn, any intimations of control and stability are flickering moments in which we find ourselves observing things accidentally align and synchronize. We begin to see no difference between what it is to control and what it is to relinquish control; we perceive non-action as action, and see a dearth of agency as a sense of agency. With this notion, landscape architecture as a modern discipline can be framed as speculative ecology, rooted in a non-access philosophy and a speculative ontology. It claims that any form of knowing, as well as design, becomes a type of speculation. A landscape design is no longer an end product but a form of speculation about relationships (ecology) which designers wish to cultivate and maintain.

To conclude, this book proposes a mode of environmental thinking and practice that lives within diverse frameworks. A shared and co-produced future relies on a plurality of knowing and thinking, and those may conflict and disagree with each other. Instead of seeing difference as motivation toward a synthesized framework, a shared future requires us to learn to live with and make efforts to attune to different frameworks around us—both human and nonhuman. A shared future may not be about increased communication, understanding, and interconnectedness, which provide enough motivation for human control and expectations. Instead, this book suggests an ontological and epistemological shift, developing a sort of cybernetic thinking by accepting that not all beings communicate, and they are therefore uncontrollable. After all, posthumanism is very much about humans, but it is also about a sense of descent rather than progress. It is about accepting the limited modes of human knowing and thinking, and then speculating on different ways to maintain dynamic living relations with others. It is about a mode of co-existence in a non-communicative and uncontrollable wild world.

Notes

1 Marx, *The Machine in the Garden: Technology and the Pastoral Ideal in America.*
2 Dayoub, Dunbabin, and Corke, "Robotic Detection and Tracking of Crown-Of-Thorns Starfish."
3 Bowes et al., "Flood Mitigation in Coastal Urban Catchments Using Real-Time Stormwater Infrastructure Control and Reinforcement Learning"; Saliba et al., "Deep Reinforcement Learning with Uncertain Data for Real-Time Stormwater System Control and Flood Mitigation"; Sadler et al., "Exploring Real-Time Control of Stormwater Systems for Mitigating Flood Risk Due to Sea Level Rise."
4 Mattern, "A City Is Not a Computer."
5 Hui, "Machine and Ecology," 63.
6 Turing, "Computing Machinery and Intelligence."
7 Silver et al., "Mastering the Game of Go without Human Knowledge."
8 Vinyals et al., "Grandmaster Level in StarCraft II Using Multi-Agent Reinforcement Learning."
9 The AlphaStar Team, "AlphaStar." https://deepmind.google/discover/blog/alphastar-mastering-the-real-time-strategy-game-starcraft-ii/. Last accessed: April 29, 2024.
10 Estrada, "Towards Sentience."
11 Gabrys, *Program Earth*, 272.
12 Heylighen and Joslyn, "Cybernetics and Second-Order Cybernetics."
13 Galison, "The Ontology of the Enemy."
14 Galison.
15 Galison, 232.
16 Lystra, "McHarg's Entropy, Halprin's Chance."
17 Halprin, *The RSVP Cycles*, 19.

18 Halprin, 4.
19 Lystra, "McHarg's Entropy, Halprin's Chance."
20 Hayles, *How We Became Posthuman.*
21 Hayles, 132.
22 Heylighen and Joslyn, "Cybernetics and Second-Order Cybernetics."
23 Hayles, *How We Became Posthuman;* Wolfe, *What Is Posthumanism?*
24 Foerster, *Cybernetics of Cybernetics.*
25 Heylighen and Joslyn, "Cybernetics and Second-Order Cybernetics."
26 Maturana and Varela, *Autopoiesis and Cognition.*
27 Lettvin et al., "What the Frog's Eye Tells the Frog's Brain."
28 Maturana and Varela, *Autopoiesis and Cognition*, 48.
29 Hayles, *How We Became Posthuman.*
30 Wiener, *The Human Use of Human Beings*, 17.
31 Hayles, *How We Became Posthuman.*
32 Heylighen and Joslyn, "Cybernetics and Second-Order Cybernetics."
33 Kuhn, *The Structure of Scientific Revolutions: 50th Anniversary Edition.*
34 Barthes, "The Death of the Author."
35 Foucault, *The Order of Things: An Archaeology of the Human Sciences*, 422.
36 Reed and Lister, *Projective Ecologies*; Reed and Lister, "Ecology and Design."
37 Wolfe, *What Is Posthumanism?*
38 Harman, "Object-Oriented Ontology," 405.
39 Young, "On Correlationism and the Philosophy of (Human) Access," 43.
40 Young.
41 Young, 50.
42 Nye, "Technologies of Landscape"; Reuss and Cutcliffe, *The Illusory Boundary*; Lee and Helphand, *Technology and the Garden.*
43 Jullien, *Living Off Landscape.*
44 Czerniak, *Downsview Park Toronto*; Steiner, "The Ghost of Ian McHarg."
45 Cantrell and Holzman, *Responsive Landscapes*, 254.
46 Davis, "Landscapes and Instruments," 305.
47 Hui, "Machine and Ecology."
48 Hui, 63.
49 Hui.
50 Young, "On Correlationism and the Philosophy of (Human) Access," 51.

Bibliography

Barthes, Roland. "The Death of the Author." In *Image-Music-Text*, translated by Stephen Heath, 142–48. London: Fontana, 1977.

Bowes, Benjamin D., Arash Tavakoli, Cheng Wang, Arsalan Heydarian, Madhur Behl, Peter A. Beling, and Jonathan L. Goodall. "Flood Mitigation in Coastal Urban Catchments Using Real-Time Stormwater Infrastructure Control and Reinforcement Learning." *Journal of Hydroinformatics*, 2020. https://doi.org/10.2166/hydro.2020.080.

Cantrell, Bradley, and Justine Holzman. *Responsive Landscapes: Strategies for Responsive Technologies in Landscape Architecture*. New York: Routledge, 2015.

Czerniak, Julia, ed. *Downsview Park Toronto*. CASE Series (Prestel Verlag). Munich; London: Prestel, 2001.

Davis, Brian. "Landscapes and Instruments." *Landscape Journal* 32, no. 2 (2013): 293–308.

Dayoub, Feras, Matthew Dunbabin, and Peter Corke. "Robotic Detection and Tracking of Crown-Of-Thorns Starfish." In *ARC Centre of Excellence for Robotic*

Vision; Faculty of Science and Technology; Institute for Future Environments. Hamburg, Germany, 2015. https://eprints.qut.edu.au/85974/.

Estrada, Leif. "Towards Sentience." In *Codify: Parametric and Computational Design in Landscape Architecture*, edited by Bradley Cantrell and Adam Mekies, 279–88. Milton Park; Abingdon; Oxon; New York: Routledge, 2018.

Foerster, Heinz von. *Cybernetics of Cybernetics, The Control of Control and the Communication of Communication*. Minneapolis, MN: Carl-Auer, 1974.

Foucault, Michel. *The Order of Things: An Archaeology of the Human Sciences*. 1st American ed. World of Man. New York: Pantheon Books, 1970.

Gabrys, Jennifer. *Program Earth: Environmental Sensing Technology and the Making of a Computational Planet*. Electronic Mediations. Minneapolis: University of Minnesota Press, 2016.

Galison, Peter. "The Ontology of the Enemy: Norbert Wiener and the Cybernetic Vision." *Critical Inquiry* 21, no. 1 (1994): 228–66.

Halprin, Lawrence. *The RSVP Cycles: Creative Processes in the Human Environment*, 1st ed. New York: George Braziller, 1970.

Harman, Graham. "Object-Oriented Ontology." In *The Palgrave Handbook of Posthumanism in Film and Television*, 401–9. London: Palgrave Macmillan, 2015. https://doi.org/10.1057/9781137430328_40.

Hayles, Katherine. *How We Became Posthuman: Virtual Bodies in Cybernetics, Literature, and Informatics*. Chicago, IL: University of Chicago Press, 1999.

Heylighen, Francis, and Cliff Joslyn. "Cybernetics and Second-Order Cybernetics." In *Encyclopedia of Physical Science and Technology*, 155–69. Amsterdam: Elsevier, 2003. https://doi.org/10.1016/B0-12-227410-5/00161-7.

Hui, Yuk. "Machine and Ecology." *Angelaki* 25, no. 4 (July 3, 2020): 54–66. https://doi.org/10.1080/0969725X.2020.1790835.

Jullien, François. *Living Off Landscape: Or the Unthought-of in Reason*. Global aesthetic research. London; New York: Rowman & Littlefield International, 2018.

Kuhn, Thomas S. *The Structure of Scientific Revolutions: 50th Anniversary Edition*. Chicago, IL: University of Chicago Press, 2012.

Lee, Michael G., and Kenneth I. Helphand, eds. *Technology and the Garden*. Dumbarton Oaks Colloquium on the History of Landscape Architecture, XXXV. Washington, DC: Dumbarton Oaks Research Library and Collection, 2014.

Lettvin, J., H. Maturana, W. McCulloch, and W. Pitts. "What the Frog's Eye Tells the Frog's Brain." *Proceedings of the IRE* 47, no. 11 (November 1959): 1940–51. https://doi.org/10.1109/JRPROC.1959.287207.

Lystra, Margot. "McHarg's Entropy, Halprin's Chance: Representations of Cybernetic Change in 1960s Landscape Architecture." *Studies in the History of Gardens & Designed Landscapes* 34, no. 1 (January 2, 2014): 71–84. https://doi.org/10.1080/14601176.2013.850313.

Marx, Leo. *The Machine in the Garden: Technology and the Pastoral Ideal in America*. New York: Oxford University Press, 1964.

Mattern, Shannon. "A City Is Not a Computer." *Places Journal*, February 7, 2017. https://doi.org/10.22269/170207.

Maturana, H. R., and F. J. Varela. *Autopoiesis and Cognition: The Realization of the Living*. Boston Studies in the Philosophy and History of Science. Cham: Springer Netherlands, 1980. https://doi.org/10.1007/978-94-009-8947-4.

Nye, David E. "Technologies of Landscape." In *Technologies of Landscape: From Reaping to Recycling*, edited by David E Nye, 3–17. Amherst: University of Massachusetts Press, 1999.

Reed, Chris, and Nina-Marie Lister. "Ecology and Design: Parallel Genealogies." *Places Journal*, April 14, 2014. https://doi.org/10.22269/140414.

Reed, Chris, and N.M.E. Lister. *Projective Ecologies*. Actar D, 2014.

Reuss, Martin, and Stephen H Cutcliffe, eds. *The Illusory Boundary: Environment and Technology in History*. Charlottesville; London: University of Virginia Press, 2010.

Sadler, Jeffrey M., Jonathan L. Goodall, Madhur Behl, Benjamin D. Bowes, and Mohamed M. Morsy. "Exploring Real-Time Control of Stormwater Systems for Mitigating Flood Risk Due to Sea Level Rise." *Journal of Hydrology* 583 (April 1, 2020): 124571. https://doi.org/10.1016/j.jhydrol.2020.124571.

Saliba, Sami M., Benjamin D. Bowes, Stephen Adams, Peter A. Beling, and Jonathan L. Goodall. "Deep Reinforcement Learning with Uncertain Data for Real-Time Stormwater System Control and Flood Mitigation." *Water* 12, no. 11 (November 2020): 3222. https://doi.org/10.3390/w12113222.

Silver, David, Julian Schrittwieser, Karen Simonyan, Ioannis Antonoglou, Aja Huang, Arthur Guez, Thomas Hubert, et al. "Mastering the Game of Go without Human Knowledge." *Nature* 550, no. 7676 (October 2017): 354–59. https://doi.org/10.1038/nature24270.

Steiner, Frederick R. "The Ghost of Ian McHarg." *Log*, no. 13/14 (2008): 147–51.

The AlphaStar Team. "AlphaStar: Grandmaster level in StarCraft II using multi-agent reinforcement learning." Deepmind, October 30, 2019. https://deepmind.com/blog/article/alphastar-mastering-real-time-strategy-game-starcraft-ii.

Turing, A. M. "Computing Machinery and Intelligence." *Mind, New Series* 59, no. 236 (1950): 433–60.

Vinyals, Oriol, Igor Babuschkin, Wojciech M. Czarnecki, Michaël Mathieu, Andrew Dudzik, Junyoung Chung, David H. Choi, et al. "Grandmaster Level in StarCraft II Using Multi-Agent Reinforcement Learning." *Nature* 575, no. 7782 (November 2019): 350–54. https://doi.org/10.1038/s41586-019-1724-z.

Wiener, Norbert. *The Human Use of Human Beings: Cybernetics and Society*, New edn. London: Free Association Books, 1989.

Wolfe, Cary. *What Is Posthumanism?* Posthumanities Series. Minneapolis: University of Minnesota Press, 2010.

Young, Niki. "On Correlationism and the Philosophy of (Human) Access: Meillassoux and Harman." *Open Philosophy* 3, no. 1 (January 1, 2020): 42–52. https://doi.org/10.1515/opphil-2020-0003.

PART I

The New Machine in the New Garden

1

A TRANSFORMATION FORMULA

In his seminal book *The Machine in the Garden*, American historian Leo Marx rendered the image of a train whistle's long shriek disturbing the serenity of an eighteenth-century American village.[1] The progressive yet destructive power symbolized by the machine was at odds with an ideal lifestyle symbolized by the garden. Many historians like Marx have posited two models—one on the order of nature, and the other on technology—against each other.

Today, Marx's garden-machine motif assumes different manifestations. Along with the RangerBot, FarmBot, and other robots in "gardens," we find ourselves in all sorts of environments that are, at least, partly machine-constructed. For example, in South Florida, numerous sensing stations have been installed across bodies of water, generating real-time hydrological and water quality data utilized to build water system simulation models. The South Florida Water Management Model is one which organizations and agencies use in order to analyze operational changes to the water system and make informed management decisions. Thousands of miles of engineered canals and pipes are carved into Florida's landscape, and water control infrastructures such as basins, spillways, weir gates, pumps, dams, and locks are strategically placed along the waterways. These are the actuators within the system that directly influence hydrological patterns in South Florida.

The Everglades restoration project in Florida exemplifies an adaptive management framework, the best practice in environmental management discourse. Adaptive management may be understood as a decision-making method through learning by doing; this approach is achieved by intensive monitoring and actuating processes. Established in 1949, the South Florida Water Management District (SFWMD) is a regional governmental agency

DOI: 10.4324/9781003320852-3

that manages water resources in the southern half of Florida, including the Everglades, a 1.5-million-acre wetland that sustains numerous wildlife species. This ultimate "wilderness" is, in fact, a highly maintained area. The amount of water and its hydrological patterns are carefully calculated and controlled, using simulation models based on real-time hydrological data, weather forecasts, climate modeling, and historical data.

The South Florida water management project is only one instance of numerous efforts to use cyberphysical systems to manage environmental processes that we largely consider natural. The recent discourse regarding smart cities has fueled cybernetic imagination across social sectors. Many cities are retrofitting their stormwater systems with controllable and smart actuators provided by emerging companies, such as Opti and EmNet, that specialize in live environmental monitoring and real-time control solutions. We live in a cybernetic environment where machines become part of the environment, constructing what we thought of as nature.

We possess the urge to use the word "nature" to denote the *nonhuman realm*, free from artificial and technological disturbances. Yet an outdated mental model—the yearning to draw a boundary between nature and technology—cannot describe a paradoxical reality. Many environmental and technology historians have argued that this boundary is more apparent in our minds than in the environment, and that such a boundary is both deceptive and misleading, particularly when we pour culturally specific values into the imagined dichotomy.[2]

Despite the intellectual critics of binary thinking, today's environmental practices are awash with dualisms, such as artificial/natural, technological/environmental, and gray/green. This binary thinking gives rise to single visions in understanding what "nature" should be and what "technology" should do. When the media describes the High Line in Manhattan as "wild in the city," they ignore the irony that the High Line's construction involved the destruction of the novel ecology of an abandoned railway, a habitat for nonhuman species that did not fit within the designers' conceptualization of "nature" and "wild." With climate change, many ecosystem types will disappear under the changing climate patterns, but new types will emerge.

Within a framework that articulates a clear division between "technology" and "nature," we can conceptualize emerging technologies, such as AI and robotics, either as destructive forces against "nature" or as tools to extend human control of "nature." Since 2021, I have led a team of computer scientists and architects to establish a research laboratory within the City College of New York, called the Ecological Laboratory for Urban Agriculture (ELUA), researching robotic-assisted urban food gardens. The ELUA team saw potential in their robot-assisted urban garden as a viable approach for climate-sensitive, small-scale urban community gardens. With this anticipation, the team began to conduct outreach efforts to identify potential community partners. However, in an early attempt, I received replies that expressed a sense of skepticism toward robots claiming their community gardens as "people-powered."

The term "people-powered" represents a sentiment many people have toward bringing robotics into community gardens. "Will robots someday replace humans?" is the underlying question. Below is a quote from a potential collaborator and their concerns about the robots, called FarmBot, used in ELUA:

> FarmBot is an interesting technology, though [it] has been ridiculed in social media, rightfully so, I think, for everything from taking the joy out of urban agriculture to de-skilling the practice and eliminating the most important co-benefits of social cohesion, community development, intergenerational knowledge sharing.

The underlying issue with this type of comment is that when people speak of urban agriculture, the first and only model imagined is community gardens managed by human labor. The critique or ridicule of attempts to bring robots into community gardens is often on the basis that technological advancement is against nature and the virtue of human labor. In other words, a "garden" symbolizing a piece of "nature" improved by human labor is the unstated premise for the moral reasonings of today's eco-gentrification narratives. In contrast, technological development, often fueled by big corporates responsible for many environmental and social problems in the first place, could symbolize values at odds with the virtue of "community gardens." The garden-machine motif persists.

With the boundary in mind, our thinking tends to pick a convenient route and articulate a simplified relationship within the environment: *humans use technologies to transform pristine nature and produce habitable landscapes* (Figure 1.1). Nature and technology are conceptualized as two kingdoms of force: the former provides limitless resources while the latter serves as an effective means to extract resources for human use.

In this formula, we can observe a series of problems, which I will articulate in three aspects. First, it ignores nonhumans and their agency in meaningfully influencing the environment, thus rendering the nonhuman realm invisible.

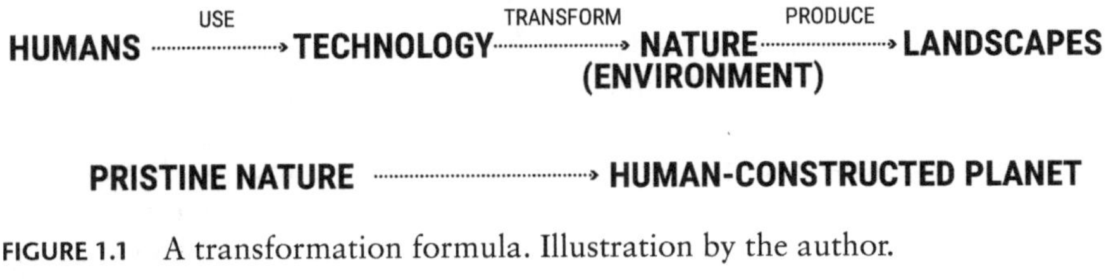

FIGURE 1.1 A transformation formula. Illustration by the author.

This formula places nonhumans and their habitat into a single basket labeled "nature," and overlooks the intrinsic complexity of the nonhuman realm. A family of beavers may build a dam that, on the one hand, serves as a shelter for them to hide from predators and, on the other hand, controls local hydrology to their benefit, such as improved food sources and a relatively stable environment. Their intervention profoundly transforms and influences the local ecosystem. Beaver dams are a classic example of "niche construction" in ecology stories, where organisms alter their local environment to support their survival, just as humans construct villages and cities that provide stable environments in which we thrive. Nevertheless, within the transformation formula, a beaver family and its ability to alter the environment are generalized as "nature," subject to human use.

The second aspect is deeply related to the first. Our ignorance of the nonhuman realm is partially due to the anthropocentrism embedded in this formula, which assumes that technology is the means by which we *humans* extend our control. However, to a certain extent, a beaver dam is technology which beavers have developed to control their environment. Again, the boundary between technology and nature is challenged by reality. We would not perceive beaver dams as technological artifacts or systems, because our concept of technology is built around imagined human agency. This simplistic formula essentially reinforces a single view of "technology" and denies "technodiversity"—the different ways a technical framework can be interpreted.[3]

Third, since the formula is generated from a human-centered perspective, it inevitably implies an apparent directionality (observe the arrows in Figure 1.1). Their unidirectionality establishes an imagined origin—an untouched, pristine nature—and the inevitable end: a human-constructed Earth. The formula articulates a seemingly inescapable process of transformation, and simultaneously constructs an image of humanity as the source of agency to initiate changes by wielding technologies. Moreover, this unavoidable transformation gives rise to the moral imperatives for the conservation and preservation narratives that have risen since the mid-twentieth century; today's mainstream environmental practices cannot succeed without first accepting the process of transformation, and then attempting to undo its errors, such as environmental degradation and pollution. Green technologies are used to recover and reconstruct a "pristine nature." However, the image of a pristine nature imposes a set of culturally specific views of nonhuman realms and a limited understanding of possible relationships, causing ongoing social and political conflicts. Recent years' political debates in the West have dealt with economic growth and plans to combat climate change, disputes between the working class and environmentalists' efforts to protect nature, and long-term contentions between Indigenous people's homelands and America's wilderness.

This transformation formula simplifies the relationship between humans, nature, and technology. Today's environmental discourse is trapped in the language of usership and stewardship. It subjugates the nonhuman realm as resources and protects it for further human extraction, ignoring the intrinsic value of nonhumans.

The transformation formula tends to be contagious. For example, one author in *The Illusory Boundary* began by recognizing the entanglement of nature and technology, yet, in the end, argued that "[t]echnology melds together the proverbial quilt of relationships between humans and nature, and out of the process emerges landscapes."[4] This quote is merely one more way to articulate the transformation formula.

We may find this formula in many contemporary environmental narratives, such as sustainability and Anthropocene. "Sustainability" is a compromise between transformation and recovery. Ironically, it reinforces and justifies further exploiting people, land, and nonhumans outside the mainstream transformation formula and recovery narratives. Anthropocene, a proposed geological epoch in which human activity has become a significant factor in altering planet Earth, pushes the formula to an extreme: we have finished transforming pristine nature, and what is left is a human-constructed Earth.

These examples illustrate how deeply our language is trapped in this mode of thinking. Even though we may want to acknowledge the complexity of the environment, we lack the vocabulary to discuss agency and its effects in non-categorical terms. It is not difficult to recognize the illusory boundary, but it is challenging to articulate the entanglement in reality, using the categorical terms which we want to avoid in the first place. Is there a model, and are there vocabularies and concepts other than the transformation formula, which we can use to understand the environment?

Environtech?

Since the early twenty-first century, scholars in the fields of environmental history and the history of technology have begun to realize that they had been studying the same historical subject.[5] In *The Illusory Boundary*, scholars provided a historical account for the entanglement of nature and technology with rich examples.[6] One essay, titled "Where Does Nature End and Culture Begin?" traced two bodies of research—environmental history and the history of technology—and posited that

> [t]he trend toward a more constructionist approach to the history of technology and away from declensions narratives in environmental history, well underway by the mid-1990s, has since led to scholarship more inclined to wrestle with the messy entanglement of technology and nature.[7]

This convergence has given rise to an "environtech" approach to challenge a traditional nature-technology dualism in historical research. However, "environtech" possesses a significant shortcoming. To see "convergence," one must accept an initial, imagined "separation." The analytical framework successfully articulated how the subjects of nature and technology are entangled in reality, but failed to explain why they had been articulated as separate realms. This acceptance of the two established realms can be identified by noting the titles of the anthology's essays: "Can Nature Improve Technology?," "The City as an Artifact of Technology and the Environment," "Where Does Nature End and Culture Begin?" These essays ask essential questions, but deploying those terms results in a dearth of complexity, and ironically reinforces the notion that "nature" and "technology" belong to two separate realms.

"Environtech" represents a popular approach in contemporary environmental design discourse through combination. Coupling, cyborg, and hybrid are popular terms, and the trend to combine "green" infrastructures with "gray" infrastructures, and engineering solutions with "nature-based" solutions continue to gain currency. Paradoxically attempting to join the two realms further articulates the boundary itself, by unwittingly accepting that they were separated in the first place. Dissolving the boundary does not necessarily prove the boundary illusory; to argue that a boundary is illusory, one must question why the dichotomy existed in the first place. The "environtech" framework revealed a paradigm shift in environmental history and the history of technology, by articulating the convergence of the two fields, but it fell short of illustrating what the new paradigm should look like, in ignoring why the boundary exists in the first place.

By reinforcing the two realms, "environtech" created a blind spot in its analytical framework, oversimplifying, if not overlooking, another category of research represented by Marx's *The Machine in the Garden*. Unlike other technology historians, Marx was not interested in the role of technology in culture and how culture shapes it; neither was he attentive to the social aspect of nature, like many environmental historians. Marx's work should represent a body of research focused on the blind spot itself, asking direct questions as to why nature and technology have been historically articulated as two kingdoms of forces. Yet because it operates in the blind spot, its value has been overlooked within the "environtech" framework. His work sat awkwardly in both environmental history and the history of technology as "rich intellectual legacies" for historians in both fields to draw upon.[8]

A Human-Technology-Nature Triad

Instead of unifying the two kingdoms, yet ironically accepting the boundary, one can bypass this dualist thinking by remapping the relationships between the concepts of nature, technology, human, and landscape. Over the past few years, a body of work has emerged in environmental humanities—environmental history, political ecology, feminist/queer/indigenous science, and technology studies. They shed light on a different conceptual framework for understanding the environment. Within this model, the environment can be articulated as a result of the co-production of many different agents, including humans, machines, animals, plants, ecological processes, and land-use policies.

To trace the development of these ideas, I rely on a human-nature-technology triad as an analytical framework (Figure 1.2). This triad represents a roadmap for the development of these ideas from the twentieth century to the early twenty-first. The ideas have gelled over time, from early simplistic and essentialist views to more complex and developmental understandings of the relationships. Toward the early twenty-first century, the ideas eventually converged into a co-production and co-evolution model.

In the following three chapters, I will focus on each edge separately. Each represents a body of work that addresses certain aspects of the evolution of ideas. I will also highlight the environmentalism that expanded in the 1960s and became part of contemporary culture. Environmental values have served as an undergirding, and provided a dynamic theme, for the development of these ideas.

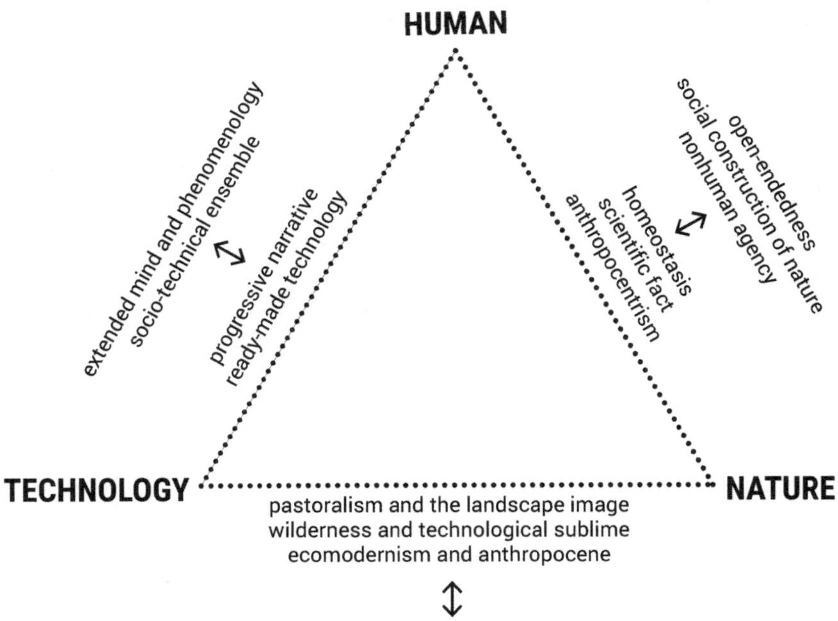

FIGURE 1.2 Human-nature-technology triad. Illustration by the author.

Notes

1 Marx, *The Machine in the Garden: Technology and the Pastoral Ideal in America.*
2 Reuss and Cutcliffe, *The Illusory Boundary: Environment and Technology in History.*
3 Hui, "Machine and Ecology."
4 Williams, "Understanding the Place of Humans in Nature," 19.
5 Reuss and Cutcliffe, *The Illusory Boundary.*
6 It should be noted that "nature" and "environment" were used interchangeably throughout the anthology. As noted by Raymond Williams in *Keywords*, using "the environment" to denote "nature" cannot be separated from the environmental movement, since the 1960s. The anthology would run the risk of not problematizing environmentalism itself, which has been critically reflected by many in the 1990s, and is still contested ground today.
7 Gorman and Mendelsohn, "Where Does Nature End and Culture Begin? Converging Themes in the History of Technology and Environmental History," 277.
8 Gorman and Mendelsohn, 275.

Bibliography

Gorman, Hugh S., and Betsy Mendelsohn. "Where Does Nature End and Culture Begin? Converging Themes in the History of Technology and Environmental History." In *The Illusory Boundary: Environment and Technology in History*, edited by Martin Reuss and Stephen H Cutcliffe, 225–90. Charlottesville, London: University of Virginia Press, 2010.

Hui, Yuk. "Machine and Ecology." *Angelaki* 25, no. 4 (July 3, 2020): 54–66. https://doi.org/10.1080/0969725X.2020.1790835.

James, C. Williams. "Understanding the Place of Humans in Nature." In *The Illusory Boundary: Environment and Technology in History*, edited by Martin Reuss and Stephen H Cutcliffe, 9–25. Charlottesville, London: University of Virginia Press, 2010.

Marx, Leo. *The Machine in the Garden: Technology and the Pastoral Ideal in America.* New York: Oxford University Press, 1964.

Reuss, Martin, and Stephen H Cutcliffe, eds. *The Illusory Boundary: Environment and Technology in History.* Charlottesville, London: University of Virginia Press, 2010.

Williams, Raymond. *Keywords: A Vocabulary of Culture and Society.* Oxford: Oxford University Press, 2014.

2

THE TECHNOLOGY-NATURE EDGE

From Pastoralism to Anthropocene

Human, nature, technology, and landscape are four elements within the transformation formula. What is the function of each element, and what are their relationships? Influential historians, including Leo Marx and David Nye, have articulated technology and nature as two kingdoms of force, and the environmental narratives in North America have been about ways to reconcile the tension between them. Tracing the nature-technology edge, I argue that the environmental narratives in North America can be categorized into three themes: (1) *pastoralism and garden-machine motif*, (2) *wilderness and technological sublime*, and (3) *Anthropocene and ecomodernism*. Even though certain ideas, such as pastoralism, date to eighteenth-century colonial America, their legacy and influence still exist. Together, these themes help form contemporary mainstream environmental narratives.

Pastoralism and Garden-Machine Motif

On May 11, 2021, former US President Donald Trump posted on his website, "From the desk of Donald J. Trump,"

> Congratulations to Martha's Vineyard in Massachusetts for the privilege they will have in looking at *massive windmills* that have been approved by the Biden Administration…as part of an extraordinarily large wind farm. Wind is an incredibly expensive form of energy that kills birds, affects the sea, *ruins the landscape,* and creates disasters for navigation….[1]

Massive windmills ruin the landscape—a perfect example of the garden-machine motif. On July 29, 2020, the Trump administration repealed

DOI: 10.4324/9781003320852-4

President Obama's Affirmatively Furthering Fair Housing Rule (AFFH), which was intended to combat housing discrimination, on the basis that it would force the construction of low-income homes in suburban areas, ruining the "suburban lifestyle." Despite the misinterpretation of the AFFH, there is clear pastoralism in Trump's narrative,

> I am happy to inform all of the people living their *Suburban Lifestyle Dream* that you will no longer be *bothered* or financially *hurt* by having low income housing built in your neighborhood... I have rescinded the Obama-Biden AFFH Rule. Enjoy![2]

If pastoralism is a form of escapism—America as a retreat from eighteenth-century Europe—then suburbanization in post-war America was an escape within an escape. Streetcars and automobiles facilitated suburban development, and American landscapes rendered an image of clean-cut lawns with single-family houses outside busy metropolises.

The conflict which many perceive between nature and technology may be generalized as a garden-machine motif, with pastoralism at its heart. Pastoralism is a sentiment across culture and time—it is the gesture of humans moving away from the "artificial" world, or "another of our many vehicles of escape from reality."[3] American pastoralism, intertwined with colonialism, attempts to reconcile two kingdoms of force—nature and technology—with a second creation, or technological creation, narrative. By telling stories around technological artifacts, from axes to water mills, dams, and railroads, early colonists constructed a history of America as God's nature transformed by human power. These narratives "naturalized the technological transformation of the United States so that it seemed an inevitable and harmonious process leading to a second creation that was implicit in the structure of the world."[4]

Early colonists conceptualized America as an "uninhabited island," an "unspoiled terrain" for a pastoral utopia.[5] Toward the end of the eighteenth century, this pastoral ideal was formalized, fully articulated, and finally realized, thanks to Thomas Jefferson, who wanted America to became a yeoman republic free from European feudal oppression, and machines played an indispensable role. According to Marx, it was Jefferson's belief that "once the machine is removed from the dark, crowded, grimy cities of Europe... it will blend harmoniously into the open countryside of his native land."[6] Pastoralism in America did not lead to the total removal of machines; instead, mechanical power was reconciled as a means to achieve a pastoral end. Jefferson "envisages it [machine power] turning mill wheels, moving ships upriver, and, all in all, helping to transform a wilderness into a society of the middle landscape."[7] Here, the image of the middle landscape is not free from machine and progressive change; on the contrary, it is "nature improved" with the help of machines, "an ideal fusion of nature with art."[8] Transformative power was rendered both harmless and necessary in early America. The pastoral image reconciles two sources of power in the transformation formula; thus, technology and nature find balance in the middle landscape.

If nature and technology create a balance within pastoralism, why is this an unsettling image? We must analyze the term "landscape" and its role in reconciling the two concepts. The term *landscape* was not always associated with aesthetic views, and its original meaning was surprisingly akin to that of "technology." Historians note that the root of "landscape" is the German word *Landschaft*, which emerged in 1121 to refer to the inhabitants of a legally defined zone. The term, in its original sense, was therefore tightly wed with the people working on the land. Only later was the term linked to landscape painting, in the sixteenth century; by the eighteenth century, it was fully integrated into the vocabulary of aesthetics.[9] The flourishing of landscape painting aestheticized and romanticized both land and labor. Of course, understanding landscape through vision cannot be separated from Western subject-object relationships, with the rise of science. Landscape "will always presume the *exteriority of the spectator.*"[10] To conceptualize landscape also means to create observers. Invented by Claude Lorrain, a seventeenth-century landscape painter, Claude glasses—small, portable convex mirrors that captured landscape views and gave them a painterly quality—were beloved by eighteenth-century European tourists. When observers see a landscape painting as outsiders who do not work the land, they naturalize those who work and make the landscape part of the view. Labor is rendered harmonious with its natural setting as part of the serenity. Where labor was aestheticized, landscape "had lost its associations with work or with direct involvement in the creation of social space."[11] Landscape is reduced to a static view that freezes time and renders an everlasting image of happiness.

One caveat is that *landscape* refers here to what was portrayed as a cultural practice instead of the specific usage of the term in the discipline of landscape architecture. Theory and practice in contemporary landscape architecture have already bypassed the myth of *landscape* as beautiful views. Many landscape architects use the term to denote complex evolving systems that may give rise to emergent forms and relations. The usage of *landscape* in the discipline is akin to the original meaning of *Landschaft* in the twelfth century. Unfortunately, this specific usage does not transcend the discipline. For most people, the term remains attached to aesthetically pleasing images tied to culturally specific values and norms.

In the transformation formula, images of the middle landscape served as an index of a pastoral lifestyle. A landscape image effectively articulates "pastoralism" by rendering vivid views in one's mind: green, rural, a house at a distance with woods in the background. Images of landscapes helped people visualize complex ideas, and, more importantly, they built a shared mental model for people to communicate those complicated ideas without words. Images of landscapes bond people together and empower those working toward "a good life." If the images of crowded eighteenth-century European cities symbolize outdated values embodied in feudalism, then the images of a green and rural middle landscape represent an idealized lifestyle of a yeoman republic in which independent farmers form the basis of republican values. The landscape images create a contrast between good and evil, thus setting up the basis for moral imperatives for transformation efforts. The images become at once a goal and an end, calling for effective means to achieve them.

However, the "green, rural image" created only deceptive stability, and the aestheticized landscape proved to be fragile against rising technological power. Although Jefferson did not envision that new mechanical power would strive against his pastoral ideal, progressive powers would rise to disturb the imagined stability, and there was a profound disconnect between the pastoral ideal and Americans' lived reality. In the early twentieth century, technology gradually became a concept to index the unprecedented human power that can initiate social change, and "the aestheticized idea of landscape, as being any attractive view, could easily seem to be the opposite of all that industrialization stood for."[12] The progressive power of technology challenges the shared landscape images in people's minds, which picture a static and homogeneous good life.

An important aspect of the perceived conflict between nature and technology is the repeated theme of a garden-machine motif in America. The aestheticized idea of landscape maintains the transformation formula intact by rendering an image of happiness, which functions both as a moral imperative and a reassurance that compels people to follow this simple formula to achieve a good life. This image evokes a goal and an end, which call for any means to achieve them. Because an aestheticized landscape freezes time, anything not included in this image becomes a threat. Landscape bound the transformation formula together but simultaneously planted the seeds of instability; whoever embarked on this formula was doomed to confront and reconcile its embedded tension, to wrestle with the garden-machine motif itself.

In the image of pastoral landscapes, specific figures of humanity—European colonists—were rendered as the sole sources of agency. As storytellers, colonists portrayed America as a new frontier, a nature's garden, an ultimate wilderness, the Creator's work; they, as the wielders of technology, had been sent to conquer and transform it. However, America was also home to the Indigenous people who, over millennia, inhabited the land and transformed it with their own "technologies"—cultural practices formed between humans and other species, "contaminating" each other's world-making projects.[13] One danger of the transformation formula is that it justifies colonizers' heinous treatment of Native Americans.

Many historians tend to overlook that an aestheticized landscape generalizes the internal complexity of the nonhuman realm. When people say "landscape," they speak to other living beings. Rendering lands as pleasing and calm images obscures the fact that they are also produced by and home to many nonhuman species, working and investing labor into the land. The real scene is chaotic, busy, and vibrant, as opposed to the serenity depicted in landscape paintings. The nonhuman realm and its agency simply exist outside the conceptual frameworks that underpin pastoralism.

Despite decades of criticism, pastoralism persists and is now intertwined with racism, white nationalism, and classism. Trump's tweets were perfect examples. The garden-machine motif now takes on new manifestations. As Marx argues, "[the garden-machine motif] is the germ... of the most final of all generalizations about America."[14] If this motif is not properly historicized, contemporary environmental narratives must perpetually confront the discontinuity between reality and the ideal mental model.

Wilderness and Technological Sublime

In the mid-twentieth century, *wilderness* was summoned as a concept to maintain the perceived balance between nature and technology. Conservation groups such as Sierra Club and The Wilderness Society have become primary advocates for protecting American wild places, by, for example, ensuring the passage of the Wilderness Act in 1964. The Act defines *wilderness* as

> an area of undeveloped Federal land retaining its primeval character and influence, without permanent improvements or human habitation, which is protected and managed so as to preserve its natural conditions and which (1) generally appears to have been affected primarily by the forces of nature, with the imprint of [hu]man's work substantially unnoticeable; (2) has outstanding opportunities for solitude or a primitive and unconfined type of recreation...and (4) may also contain ecological, geological, or other features of scientific, educational, scenic, or historical value.[15]

The concept of wilderness has two effects on the transformation formula. First, using the terms "primeval" and "primitive" freezes a "prehistorical" time as the starting point, based on which humans may make "improvements" to these "natural conditions." It veils a process of shifting baselines; conservationists saw wildness in the places many early settlers called home, and the settlers saw wildness in the lands in which Indigenous peoples had lived and transformed for millennia. Moreover, from a nonhuman perspective, there is nothing wild about "wilderness." My criticism here does not intend to underestimate preservation and conservation efforts because the unstated agenda to protect the nonhuman habitat essentially became a political goal. Yet the incentives—scientific, educational, scenic, or historical values—had nothing to do with the intrinsic values of the nonhuman world. The concept of wilderness reinforced the "pristine" state of nature by articulating an imagined origin and excluding multiple generations of humans and nonhumans from the wilderness narrative.

Second, within the wilderness narrative, human habitation—or the middle landscape—is assumed to be an "improvement"—a need to transform nature through technologies. The wilderness image provides a baseline against which progress in the second creation stories can be measured and justified.[16] "Improvement" imposes positive values on the action of transforming so that the transformation formula itself remains morally intact and compels people to act.

Wilderness in America is more than an idea; it is an ideology. In his (in)-famous 1996 essay "The Trouble with Wilderness," environmental historian William Cronon traces the development of wilderness in America, arguing that it came to embody a sheaf of moral values and cultural symbols deeply rooted in Judeo-Christian traditions, as well as in narratives of escapism, romanticism, primitivism, and nationalism.[17] Wild places have become icons for religious redemption and national renewal, and these associations have fueled the moral imperatives for the many twentieth-century conservation and preservation efforts. With such a notion of wilderness, early reconciliation narratives are replaced by a preservation and conservation narrative. As Cronon noted, "the modern environmental movement is itself a grandchild of romanticism and post-frontier ideology, which is why it is no accident that so much environmentalist discourse takes its bearings from the wilderness these intellectual movements helped create."[18]

Today's environmental discourse, including urban and landscape design, is awash with claims to reconstruct wild places in cities. Cities, as our largest technological artifacts, represent ultimate order and human control. Conserving wild places in a metropolis thus creates an extreme contrast between "nature" and "technology." As a descendant of wilderness ideology, many rewilding efforts impose a static and singular vision based on a wilderness image of a lack of human inhabitation and disturbance. This narrow conceptualization of wild places possesses many shortcomings.

First, urban wilds evoke a specific image of a dearth of human presence, or the opposite of home. However, what we refer to as wilderness may be home for others. In America, the establishment of national parks, many of which embody a sense of wilderness, involved the removal of Indigenous people from their homelands. When conservationists tried to protect the Amazon rainforest as a section of wilderness, they ignored the Indigenous groups that had for millennia inhabited and managed the forest. When the media described the High Line as "wild in the city," they overlooked the irony that construction of the High Line led to the destruction of a novel ecology of an abandoned railway, which is habitat for nonhuman species that fail to fit within the designers' category of *wild*. Today's rewilding discourse reinforces a narrow view of urban wild, based on a specific perspective from urban dwellers that ignores other perspectives on "wildness" and how it may be interpreted and defined.

Second, wilderness freezes time and ignores evolution and emergence. The narratives of wilderness often imagine a time before human interventions and attribute moral values to that time frame; environmentalists would attempt to restore places based on an ideal image, when that image is, in fact, only a fraction of the continuous unfolding of ecosystems. Contemporary landscape theory and practices have widely acknowledged that physical landscapes are never static, but are forever a "shifting mosaic," a pattern of "sporadic, repeated emergences and disappearances of different ecosystem types."[19] Especially if we consider climate change, many ecosystem types will disappear, but new types will emerge. The intensively maintained "wild places," such as the High Line, leave no room for these new types of wildness.

Third, and most importantly, wilderness enlarged the false dichotomy between human versus nature—a home here, a wild place out there. As Cronon argues, this dualism entices us to focus on protecting a distanced wild "nature," but overlook the "wildness" within and around us.[20] As a result, environmental strategies can be conceptualized only as preservation, conservation, and restoration efforts, building "natural sceneries" that look like unmanaged places to erase traces of artificiality, and protecting "leftover" places in urbanization processes as if they were isolated systems without human influence.

In a sense, the preservation and conservation of wild places reflect society's urge to rebuild an imagined equilibrium. If technology becomes the symbol of progress, then wilderness represents the existence of a "pristine nature" where people find reassurance. This may be the deepest moral imperative for contemporary urban wild discourse—to find reassurance in seeing wildlife in urban environments. Wilderness, as a concept to balance the progressive power of technology and maintain an equilibrium, ironically enlarged the gap between the two realms. The garden-machine motif escalated into a confrontation between technology and wilderness—the two opposing ends on a conceptual spectrum of structure and randomness. To reconcile the conflict, another concept was summoned—the sublime.

Charting the technological history of America, David Nye declares that sublimity found a way to reconcile the two extremes. *Sublime* is a category of aesthetic experience constructed by Enlightenment philosophers such as Edmund Burke and Immanuel Kant. Despite the nuances between different conceptualizations, sublime points to the emotion experienced when humans encounter "extreme magnitude or vastness" or contemplate "scenes that arouse terror" but are safe to observe.[21] As a philosophical object of inquiry, it explains human reasoning. According to Kant, sublime lies in the experience where humans are presented with the magnificence of nature, while still being able to comprehend it with reason and cognitive powers. As opposed to beauty, which everyone can understand, sublime belongs to those with strong sensibility and rationality.[22] Natural scenery, such as the Grand Canyon, Niagara Falls, and Yosemite, or disastrous events, such as volcanic eruptions and earthquakes, have always served as examples of images that help philosophers conceptualize sublime.

Though sublime began as a philosophical idea and an aesthetic inquiry, the American sublime became a cultural practice mixed with religion, nationalism, and technology; curiously, people have discovered sublime not only in wild nature but also in large-scale technological artifacts and systems such as the Hoover Dam, the Golden Gate Bridge, railroads, the telegraph, skyscrapers, and the Apollo Project.[23] These sublime events were meant to be experienced by masses of people because "[o]ne of the most powerful human emotions, when experienced by large groups, the sublime can weld society together."[24] In America, sublime is less about human self-cultivation in philosophers' terms and addresses more the functions of the human emotions of amazement and awe within a cultural context. It "transformed the individual's experience of immensity and awe into a belief in national greatness."[25] Here, sublime functions akin to pastoralism in terms of reconciling nature and technology. While pastoralism mixes nature and technology in the same category as the "natural" cause for the middle landscape, sublimity unifies nature and technology as a source of national pride. "New technologies become self-justifying parts of a national destiny, just as the natural sublime once undergirded the rhetoric of manifest destiny."[26]

The landscape image and its ability to reduce complex ideas into vivid mental models played an essential role in wilderness and sublimity narratives. Landscape paintings and photographs of wild places freeze a series of iconic views; they teach people the "proper" way to view the wilderness. When we speak of Yosemite, we automatically think of the images of the landscape painter Albert Bierstadt and wallpaper on a Macintosh computer screen. Curiously, the views of technological artifacts are depicted like natural scenery—the image of the Golden Gate Bridge and the Hoover Dam, for example. For the American sublime, the landscapes of nature and technology belong to the same category, as long as they evoke the same type of emotional reactions. Nature and technology are two essential elements, and a landscape image is a vehicle to deliver complex ideas; in this fashion, the fundamental structure of pastoralism finds its way into American sublimity.[27]

Narratives around wilderness and sublimity reveal a human image. Consequently, there is no wilderness alone, but "wilderness for." The concept itself creates an observer and an outsider. Yet when we speak of enjoying the great outdoors, it is, in fact, a small fraction of us—middle-class urban residents with the leisure and privilege to drive outdoors—that can enjoy the "wilderness." Sublime is how observers process their emotional reactions toward whatever they perceive as wild and alien. In essence, the Kantian sublime was an object of inquiry for human reasoning, but this image of the human—"enlightened European man"—is an essentialist definition based on strong rationality and cognitive powers. The American technological sublime is fundamentally a testimony to human agency and a celebration of those who wield technology to transform nature.

Anthropocene, Half-Earth, Ecomodernism, and Green Technology

The atomic bombs that ended World War II pushed the technological sublime to its fringe, and sublimity can no longer veil the darker side of technological development.[28] The water and air pollution that led to strong environmental policies in the 1960s exemplified the belief that environmental degradation was the direct, inevitable result of industrialization and technological progress. By the 1960s, scholars from many disciplines critiqued society's choices regarding technological development, and relationships between society and technology have gradually become an important discourse, eventually growing into a field of study. Examples include the founding of the Society of History of Technology in 1958, as well as the establishment of interdisciplinary programs across universities under the banners of Science Technology and Society (STS), Science and Technology Studies (S&TS), and Sociology of Scientific Knowledge (SSK) in the 1960s and 1970s. Most of these involved empirically based studies analyzing the complexity of socio-technological networks. These disciplinary inquiries have provided various models to conceptualize society's relationship with technology; toward the late 1980s, there was a growing emphasis on the understanding that technical artifacts and systems are socially constructed.[29]

Even though nature or the environment is not, in most cases, at the center of STS analysis, we can observe a clear involvement. One example was the influential anthology *The Social Construction of Technological Systems* (1987). The ideas developed in that anthology, such as actor-network theory (ANT), found their way outside the field and proliferated in the twenty-first-century discourse of posthumanism and political ecologies. Many writers use ANT to frame nonhuman agency.[30] Thus, STS has become a supra-disciplinary method and an important lens for analysis in contemporary academics, providing vocabularies and frameworks within which to consider the entanglement of nature and technology.

Despite this intellectual development and criticism of dualism, present-day environmental narratives still simplify the relationship between nature and technology. Many contemporary narratives rely on the notion of the

Anthropocene. The Anthropocene is a proposed geological epoch in which humans are major factors in shaping Earth's biosphere.[31] A 1990s study has found traces of lead from Greek and Roman times in ice cores taken from Greenland, suggesting that ancient lead and silver mining, and the smelting of metals, polluted the middle troposphere of the Northern Hemisphere on a hemispheric scale two millennia ago.[32] As of 2020, the global anthropogenic mass surpasses all global living biomass.[33] Thus, we are living on a human-made planet.

The Anthropocene's message is clear: What a mess we humans have made of Earth! To state bluntly that humans have become the major factor in shaping Earth would generate great incentives to mobilize different social sectors in response to the environmental crisis. Under the pressure of the idea that humans have shaped the environment on a planetary scale, narratives between technology and nature have emerged. Based on their sentiment toward technology, I would term them "downscaling" and "upscaling" approaches.

In the downscaling narratives, one ambitious proposal is the Half-Earth movement. First proposed as "Nature Needs Half" by Harvey Locke,[34] then *Half-Earth* by E.O. Wilson,[35] with further development by Dinerstein et al.,[36] this conservation effort seeks to set aside half of Earth's terrestrial biosphere as a conservation reserve. A similar idea is the planetary boundaries proposed by system and environmental scientists.[37] The intention is to set up system-process boundaries to limit human activities, since scientists believe that transgressing the boundaries would lead to non-linear and abrupt environmental change. Both ideas, one arguing for setting limits on the human realm, the other insisting on setting aside half of the planet for "nature," are based on a similar sentiment and conviction that "nature" is a system too complex to be comprehended by the human mind, and thus we had better let it be. We may see ideas such as Half-Earth and planetary boundaries as present-day continuations of wilderness narratives concerning the role of technology in the environment, viewing modern technological systems and modernization as the causes of environmental degradation.

In contrast, an upscaling narrative has also formed along the line of the Anthropocene. The argument goes like this: Since humans have become the major factor in shaping the planet, we should then take extra care of it by actively constructing an anthropogenic planet that better sustains us. This view does not seek to separate humans from nature; rather, it stems from growing ecological and philosophical reasonings that humans have always been part of "nature," and human factors are an intrinsic part of ecosystems. Underpinning the upscaling narratives is the challenge to present-day ecological and sustainable ideologies, with contemporary technological realities and ethical predicaments. In a sense, the world population relies on intensive technological systems to survive. It is unrealistic to retreat to traditional organic farming or dependence on harvesting biomass for energy; teaching ecology in that way is "pedagogical malpractice."[38] Instead, the upscaling narratives seek to influence a paradigm shift, from "natural systems with humans disturbing them" to a new paradigm of "societies sustaining an anthropogenic biosphere."[39] This line of reasoning further develops into what is known as ecomodernism, which challenges contemporary mainstream environmental discourse awash with environmental protection and recovery narratives. It believes in "humanity's extraordinary powers in service of creating a good Anthropocene."[40] What distinguishes ecomodernists from Half-Earth proponents is that the former believe that protecting only half of Earth is "not nearly enough" and is only the beginning of a more integral human enterprise to conserve biodiversity.[41] A similar view is the Earth Systems Engineering and Management (ESEM) approach.[42] This involves an application of systems thinking on a planetary scale, understanding Earth as coupled human-nature systems—a product of human design. Rather than assuming a high degree of knowledge and certainty about environmental systems, ESEM will be in constant dialogue with them through an adaptive management framework.

Present-day environmental narratives combine upscaling and downscaling narratives as two major ingredients, mixed with various culturally specific values and moral reasoning. Yet, a common thread is that "green and sustainable technologies" are envisioned as necessary for human survival in the Anthropocene. These narratives use the catch-all terms "green" and "sustainable" to infuse technology with complex environmental values. By alleviating, or even eradicating, its perceived conflicts with nature, the notion of "green technology" reassures users regarding the need for technology to maintain the middle landscape image. Green technology sends a message: not only can technology construct middle landscapes, but it can also decouple human impact from extrinsic "nature," creating a more efficient human realm.

These views surrounding "green technology" reinforce the transformation formula by promoting a recovery narrative: wrongdoings in the transformation process may be undone. From this vantage, no fundamental difference exists between contemporary environmental narratives and the arguments advanced by early thinkers such as Thomas Jefferson and his contemporary Tench Coxe, who also viewed technological power as "natural means of happiness." Despite updated concepts such as Anthropocene and Earth systems, their underlying premise, shockingly, remains the same. Pastoralism and the basic structure of the transformation formula persist.

The darker side of green technology and recovery narratives is the conflict between local and global, or a tunnel vision of the NIMBY (not-in-my-backyard) syndrome. In recent years, smart technologies and AI systems have become yet another transformative force integrated into the green technology narrative. Proponents of smart cities imagine digitally networked infrastructures as a cure to a plethora of environmental problems, by inserting sensors and chips into every object to form an Internet of Things (IoT). Innovation narratives combined with sustainable incentives give the impression that if urban infrastructures were increasingly connected by digital technologies, urban systems could function more efficiently and sustainably, and thus the environmental crisis would be solved. Developing smart and green technologies becomes the end instead of the means. "Smartness and greenness" become a metric that measures our progress in combating the climate crisis.

However, tracing the global material flow of digital technology reveals a disturbing fact. Coltan, or columbite-tantalum ore, which contains tantalum, is used to make the capacitors that are essential in smart technologies. In the early 2000s, coltan mining in the Democratic Republic of Congo was repeatedly criticized for its systematic exploitation of the local environment and people. Smart technologies that play an important role in today's sustainability narratives are likely to be built with tantalum (in places, the ore is hand-mined) in regions that do not fit within the sustainability category. Our sustainability narratives are likely built on others' disasters.[43] Electric cars, often branded as gestures toward eco-friendly and emission-free lifestyles, provide a beautiful façade that hides an ugly reality: the electricity used to charge their batteries still needs to be generated somewhere else, commonly by burning fossil fuels. In addition, their used lithium-ion batteries end up in landfills.[44] Behind the complex climate and environmental models running on supercomputers lurks a vast amount of energy consumption,[45] which is either ignored or simply treated as an essential and inevitable cost in the name of science. What is being exploited are people, places, and numerous nonhuman species, all of which exist outside the recovery narratives built around green and smart technology.

Concerning the idea of the Anthropocene, one can reach a radical conclusion: as humans, we have always lived in an anthropogenic biosphere, and what we believed and conceptualized to be nature was fashioned by those who came before us. This approach posits that humans actively construct concepts and knowledge of the world, including that of nature. When humans began to conceptualize "nature," nature had already experienced human touch. As a result, one can argue either that there was no nature at all or that nature was not only conceptually but also physically constructed by humans. Either way, this recursive reasoning does not stop one from imagining a pristine state before humans entered the stage. If "nature" denotes a prehistoric time, the Anthropocene is just one extreme condition of the transformation formula; humans have finished the transformation process, and no "nature" is left. The idea of the Anthropocene relies on and, in turn, reinforces the transformation formula by articulating a naturalized human history of how this transformation has occurred. It provides not only a starting point but an endpoint as well. It associates time with minimum and maximum conditions and presents an inevitable environmental history.

By conceptualizing the Anthropocene, a specific human image was reinforced: those who could wield progressive power and possessed enough privilege to view themselves as the source of agency. Yet this obscures the identities of those truly responsible for the environmental crises by deploying the concept of humans to encompass the entire human species. Thus, many scholars entertain the idea of *Capitalocene* as a way to the capitalist system, which compels humans to treat the environment as an enormous resource.[46] The capitalist system permits manipulation of the environment and material flow on a global scale, such as the extraction, transport, and sale of fossil fuels across the globe.

On the other hand, the Anthropocene overlooks the complexity and agency of the nonhuman realm. Humans may be a major environmental factor, but this does not mean that humans are the only species that can construct and meaningfully influence the environment. Many types of "nature" have survived and thrived after the nuclear disasters in the Chernobyl and Fukushima power plants, mutating, evolving, and slowly transforming these post-apocalyptic exclusion zones into new forms of wild places.[47] To use Cronon's words, "To think ourselves capable of causing 'the end of nature' is an act of great hubris, for it means forgetting the wildness that dwells everywhere within and around us."[48]

The Predicament of Counter-Narratives

Tracing these three themes, it is clear that nature and technology symbolize two contrasting sets of values. Technology represents progressive, transformative, and sometimes destructive power; nature implies abundance, resources, and the power to cleanse. To reconcile nature and technology is to build narratives that merge the two opposing sides and neutralize their perceived conflict. Pastoralism, sublimity, ecomodernism, and the Anthropocene have become popular themes with which to relate stories of the interplay of nature and technology in modern and contemporary environmental narratives.

David Nye argues that the dynamic themes between nature and technology in America can be generalized into three interlocking narratives: (1) second creation narrative, (2) wilderness origin, and (3) recovery narrative.[49] He argues that these narratives support each other and have become mainstream beliefs in American culture. Specifically, the second creation narrative began the story of technological transformation; natural abundance and increasing access to transformative power helped early colonists to realize a pastoral ideal in the middle landscape. When this narrative was challenged by environmental reality, such as environmental degradation and environmental injustice, people built two more narratives to support the story of transformation. The recovery narrative promises that wrongdoings in the transformation process may be undone, whereas the wilderness origin presents a mythical baseline against which the achievements of the second creation may be measured and justified. Together, these three interlocking narratives represent the mainstream belief of the dynamic relationship between nature and technology that undergirds various contemporary environmental narratives.[50]

Although the body of work presented in this chapter primarily focuses on the American context, it reveals a basic structure between human, nature, technology, and landscape, and illustrates how the transformation formula functions. Four elements can index different aspects and events in different times and locations: a stable equilibrium (aestheticized landscape image); an imagined origin (nature); a progressive power (technology); and, most importantly, the agents who wield the progressive power and exercise their transformative agency. A well-established relationship exists between these elements: those who use progressive power can initiate change to either transform the origin or recover the origin to maintain an imagined equilibrium. The origin provides a baseline against which to measure progress and also creates an image for recovery efforts. These four elements reinforce each other through environmental narratives and form a source from which various moral imperatives are generated. This is a full cycle of the transformation formula, within which contemporary mainstream environmental narratives are trapped.

There have been numerous critical reflections on the mainstream rhetoric. For example, as Nye notes, each mainstream narrative is countered by narratives that unearth an entirely different environmental history. These stories include nonhuman species, failed farmers, and Indigenous people in the second creation, in addition to human and nonhuman inhabitants in the perceived wilderness, as well as the stories of the workers whose lives are set against ecological concerns. Unlike mainstream narratives that justify each other and form an interlocking history, the counter-narratives are disconnected and sporadic, and fail to form a coherent voice to challenge the mainstream belief.[51] To abandon the mainstream narratives means "recognizing historical injustices to the first inhabitants, accepting environmental limits... [and] the loss of white entitlement to the continent."[52]

Recent years have seen numerous efforts to challenge American mainstream environmental narratives. Both the Wilderness Society and the Sierra Club have made equity, inclusion, and justice their key agendas and principles.[53] These efforts must be praised for their recognition of parallel histories along with the mainstream wilderness narrative. At least these efforts, to some degree, have recognized that the concept of wilderness was developed to favor a small fraction of humanity. However, underlying this inclusion narrative is the unchallenged conception that we should "enjoy" and "use" the nonhuman realm as a resource. Nonhumans and their intrinsic value still lie outside contemporary wilderness stories. To a certain extent, one must discard the conception of wilderness to truly embrace the wildness of the nonhuman realm.

Another example is the increased popularity of the land acknowledgment movement among North American universities and non-profit organizations in recent years. Land acknowledgment, or land recognition, is a formal, vocal statement at the beginning of events and activities, to recognize and respect the ongoing relationship between Indigenous people and the land. Many argue that acknowledgment is the first step toward social and environmental justice for Indigenous people and their cultures. However, associated with this practice is another conundrum supported by binary thinking: the recognition of indigenous cultures and people unwittingly produces a "cultural other," whose traditions and way of life are romanticized, instrumentalized, and further exploited as a way to reinforce the mainstream narrative of technological transformation. The act of acknowledgment may freeze a period of time as an origin where history started; based on this imagined origin, the technological transformation story can be told and naturalized.

Furthermore, a second irony lies in the term "indigenous," defined as "naturally occurring." Present-day environmental narratives cannot bypass a problematic conception formed in early colonial times, when colonists believed they embodied "culture" and Indigenous people formed part of "nature," waiting to be transformed. Moreover, there were numerous nonhuman inhabitants in existence before the alleged "first inhabitants" arrived, and their stories have yet to be told.

This may be the largest predicament in the creation of counter-narratives: critiques may forever be assimilated into the mainstream narrative in one way or another, without fundamentally challenging it. No matter how many counter-narratives we unearth, the well-established relationships among nature, human, technology, and landscape remain intact. As long as one freezes a period of time as an imagined origin where history starts, a tale of humanity, nature, and technology may be told through interwoven stories of technological transformation. In these stories, there will be those who conceptualize themselves as wielders of technology and those humans and nonhumans who are "transformed" as part of "nature." To genuinely challenge mainstream environmental narratives is to think beyond the transformation formula as the fundamental explanation of our relationship with the environment.

In tracing the technology-nature edge of the triad, a specific image of humans emerged. Two kingdoms of force had to be summoned by those with the privilege to conceptualize themselves as the source of agency and to wield the power of both kingdoms. The transformation formula was a convenient design to help early colonist settlers conceptualize human agency in altering the environment. In an anthropocentric framework, nature and technology are two essential concepts to describe how humans interact with their environments. Nature, technology, and landscape are all concepts to conceptualize human agency. To bypass the predicament in contemporary environmental discourse requires us to take on a posthumanist framework and rebuild narratives that do not revolve around humans. From this vantage, although historians such as Nye and Marx have introduced a predicament in contemporary environmental discourse by presenting a collection of vivid examples, unfortunately, their historical perspectives fall short of providing useful alternative models to help us overcome the predicament itself.

Tracing this body of work also reveals a deeper, yet unquestioned, mental model. These narratives begin with the premise of two kingdoms of forces; theories were then developed to mitigate the gap between the two. In other words, both nature and technology were accepted as *a priori* where forces are generated. This neglects that the terms *nature* and *technology* possess histories of their own. This gap merits further investigation along the other two edges, which I will illustrate in the next two chapters.

Notes

1 Emphasis added. The website "From the desk of Donald J. Trump," Trump's own social media when he was banned by X (Twitter), is no longer accessible, but this quote can be found in other online articles.
2 X (Twitter) @realDonaldTrump, Jul 29, 2020, emphasize added. https://twitter.com/realDonaldTrump/status/1288509568578777088. Last accessed: April 29, 2024.
3 Marx, *The Machine in the Garden: Technology and the Pastoral Ideal in America*, 10.
4 Nye, *America as Second Creation: Technology and Narratives of New Beginnings*, 6.
5 Marx, *The Machine in the Garden: Technology and the Pastoral Ideal in America*.
6 Marx, 150.
7 Marx, 150.
8 Marx, 228.
9 Nye, "Technologies of Landscape."
10 Jullien, *Living Off Landscape: Or the Unthought-of in Reason*, 9.
11 Nye, "Technologies of Landscape," 14.
12 Nye, 15.
13 Tsing, *The Mushroom at the End of the World*.
14 Marx, *The Machine in the Garden: Technology and the Pastoral Ideal in America*, 353.
15 The Wilderness Act. https://www.nps.gov/subjects/wilderness/upload/W-Act_508.pdf, Last accessed: April 29, 2024.
16 Nye, *America as Second Creation: Technology and Narratives of New Beginnings*.
17 Cronon, "The Trouble with Wilderness: Or, Getting Back to the Wrong Nature."
18 Cronon, 72.
19 Hill, "Shifting Sites," 146.
20 Cronon, "The Trouble with Wilderness: Or, Getting Back to the Wrong Nature."
21 Nye, *American Technological Sublime*, 7.
22 Kant, *Critique of Judgement*.
23 Nye, *American Technological Sublime*.
24 Nye, xiii.
25 Nye, 43.
26 Nye, 282.
27 Here, I will extend Nye's American sublimity to what I would call "nationalist sublimity." The practice of creating sublime objects and the proliferation of the images of these objects can be seen in many nations besides the US. For example,

the Three Gorges Dam in China has been instrumentalized to build national pride. Tourists' experience of the vastness of the dam becomes a form of education for nationalism and patriotism.

28 Nye, *American Technological Sublime*, 209.
29 Gorman and Mendelsohn, "Where Does Nature End and Culture Begin? Converging Themes in the History of Technology and Environmental History."
30 Such as Jones and Cloke, *Tree Cultures: The Place of Trees and Trees in Their Place*.
31 Crutzen, "Geology of Mankind"; Ellis, "Ecology in an Anthropogenic Biosphere"; Steffen, Crutzen, and McNeill, "The Anthropocene"; Steffen et al., "The Trajectory of the Anthropocene."
32 Hong et al., "Greenland Ice Evidence of Hemispheric Lead Pollution Two Millennia Ago by Greek and Roman Civilizations."
33 Elhacham et al., "Global Human-Made Mass Exceeds All Living Biomass."
34 Locke, "Nature Needs Half."
35 Wilson, *Half-Earth*.
36 Dinerstein et al., "An Ecoregion-Based Approach to Protecting Half the Terrestrial Realm."
37 Rockström et al., "Planetary Boundaries"; Steffen et al., "Planetary Boundaries."
38 Ellis, "Ecology in an Anthropogenic Biosphere."
39 Ellis.
40 "An Ecomodernist Manifesto." https://www.ecomodernism.org/. Last accessed: April 29, 2024
41 Ellis, "To Conserve Nature in the Anthropocene, Half Earth Is Not Nearly Enough."
42 Allenby, *Reconstructing Earth: Technology and Environment in the Age of Humans*.
43 Kaika, "Our Sustainability Is Someone Else's Disaster."
44 Gonçalves, "Are Electric Cars Really Eco-Friendly?"
45 Czarnul, Proficz, and Krzywaniak, "Energy-Aware High-Performance Computing."
46 *Capitalocene* was popularized by Donna Haraway and environmental historian Jason Moore, who learned the term at a seminar in 2009 in Lund, Sweden, when then-graduate student Andreas Malm first proposed the term.
47 Deryabina et al., "Long-Term Census Data Reveal Abundant Wildlife Populations at Chernobyl"; Lyons et al., "Rewilding of Fukushima's Human Evacuation Zone."
48 Cronon, "The Trouble with Wilderness: Or, Getting Back to the Wrong Nature," 89.
49 Nye, *America as Second Creation: Technology and Narratives of New Beginnings*.
50 Nye.
51 Nye.
52 Nye, 294.
53 https://www.wilderness.org/news/article/our-commitment-diversity-equity-and-inclusion; https://www.sierraclub.org/equity.

Bibliography

Allenby, Braden R. *Reconstructing Earth: Technology and Environment in the Age of Humans*. Washington, DC: Island Press, 2005.

Cronon, William. "The Trouble with Wilderness: Or, Getting Back to the Wrong Nature." In *Uncommon Ground: Rethinking the Human Place in Nature*, edited by William Cronon, 1st ed., 69–90. New York, London: W. W. Norton & Company, 1996.

Crutzen, Paul J. "Geology of Mankind." *Nature* 415, no. 6867 (January 3, 2002): 23. https://doi.org/10.1038/415023a.

Czarnul, Pawel, Jerzy Proficz, and Adam Krzywaniak. "Energy-Aware High-Performance Computing: Survey of State-of-the-Art Tools, Techniques, and Environments." Review Article. Scientific Programming. Hindawi, April 24, 2019. https://doi.org/10.1155/2019/8348791.

Deryabina, T. G., S. V. Kuchmel, L. L. Nagorskaya, T. G. Hinton, J. C. Beasley, A. Lerebours, and J. T. Smith. "Long-Term Census Data Reveal Abundant Wildlife Populations at Chernobyl." *Current Biology* 25, no. 19 (October 5, 2015): R824–26. https://doi.org/10.1016/j.cub.2015.08.017.

Dinerstein, Eric, David Olson, Anup Joshi, Carly Vynne, Neil D. Burgess, Eric Wikramanayake, Nathan Hahn, et al. "An Ecoregion-Based Approach to Protecting Half the Terrestrial Realm." *BioScience* 67, no. 6 (June 1, 2017): 534–45. https://doi.org/10.1093/biosci/bix014.

Elhacham, Emily, Liad Ben-Uri, Jonathan Grozovski, Yinon M. Bar-On, and Ron Milo. "Global Human-Made Mass Exceeds All Living Biomass." *Nature* (December 9, 2020): 1–3. https://doi.org/10.1038/s41586-020-3010-5.

Ellis, Erle C. "Ecology in an Anthropogenic Biosphere." *Ecological Monographs* 85, no. 3 (2015): 287–331. https://doi.org/10.1890/14-2274.1.

———. "To Conserve Nature in the Anthropocene, Half Earth Is Not Nearly Enough." *One Earth* 1, no. 2 (October 2019): 163–67. https://doi.org/10.1016/j.oneear.2019.10.009.

Gonçalves, André. "Are Electric Cars Really Eco-Friendly? Maybe Not as Such Much as You Think." *Youmatter* (blog), September 25, 2018. https://youmatter.world/en/are-electric-cars-eco-friendly-and-zero-emission-vehicles-26440/.

Gorman, Hugh S., and Betsy Mendelsohn. "Where Does Nature End and Culture Begin? Converging Themes in the History of Technology and Environmental History." In *The Illusory Boundary: Environment and Technology in History*, edited by Martin Reuss and Stephen H. Cutcliffe, 225–90. Charlottesville, London: University of Virginia Press, 2010.

Hill, Kristina. "Shifting Sites." In *Site Matters: Design Concepts, Histories, and Strategies*, edited by Carol Burns and Andrea Kahn, 131–55. New York: Routledge, 2015.

Hong, Sungmin, Jean-Pierre Candelone, Clair C. Patterson, and Claude F. Boutron. "Greenland Ice Evidence of Hemispheric Lead Pollution Two Millennia Ago by Greek and Roman Civilizations." *Science* 265 (September 1, 1994): 1841–43. https://doi.org/10.1126/science.265.5180.1841.

Jones, Owain, and Paul J Cloke. *Tree Cultures: The Place of Trees and Trees in Their Place*. Oxford, New York: Berg, 2002.

Jullien, François. *Living Off Landscape: Or the Unthought-of in Reason*. Global aesthetic research. London, New York: Rowman & Littlefield International, 2018.

Kaika, Maria. "Our Sustainability Is Someone Else's Disaster: Cities and the Environment." *Green European Journal*, January 5, 2018. https://www.greeneuropeanjournal.eu/our-sustainability-is-someone-elses-disaster-cities-and-the-environment/.

Kant, Immanuel. *Critique of Judgement*. Edited by Nicholas Walker. Translated by James Creed Meredith. Revised edition. Oxford ; New York: Oxford University Press, 2009.

Locke, Harvey. "Nature Needs Half: A Necessary and Hopeful New Agenda for Protected Areas." *Nature New South Wales*, September 2014. https://search.informit.org/doi/abs/10.3316/INFORMIT.689951476354495.

Lyons, Phillip C., Kei Okuda, Matthew T. Hamilton, Thomas G. Hinton, and James C. Beasley. "Rewilding of Fukushima's Human Evacuation Zone." *Frontiers in Ecology and the Environment* 18, no. 3 (2020): 127–34. https://doi.org/10.1002/fee.2149.

Marx, Leo. *The Machine in the Garden: Technology and the Pastoral Ideal in America*. New York: Oxford University Press, 1964.

Nye, David E. *America as Second Creation: Technology and Narratives of New Beginnings*. Cambridge, MA: MIT Press, 2003.

———. *American Technological Sublime*. Cambridge, MA: MIT Press, 1994.

———. "Technologies of Landscape." In *Technologies of Landscape: From Reaping to Recycling*, edited by David E Nye, 3–17. Amherst: University of Massachusetts Press, 1999.

Rockström, Johan, Will Steffen, Kevin Noone, Åsa Persson, F. Stuart III Chapin, Eric Lambin, Timothy Lenton, et al. "Planetary Boundaries: Exploring the Safe Operating Space for Humanity." *Ecology and Society* 14, no. 2 (November 18, 2009). https://doi.org/10.5751/ES-03180-140232.

Steffen, Will, Wendy Broadgate, Lisa Deutsch, Owen Gaffney, and Cornelia Ludwig. "The Trajectory of the Anthropocene: The Great Acceleration." *The Anthropocene Review* 2, no. 1 (2015): 81–98. https://doi.org/10.1177/2053019614564785.

Steffen, Will, Paul J. Crutzen, and John R. McNeill. "The Anthropocene: Are Humans Now Overwhelming the Great Forces of Nature." *AMBIO: A Journal of the Human Environment* 36, no. 8 (2007): 614–21. https://doi.org/10.1579/0044-7447(2007)36[614:TAAHNO]2.0.CO;2.

Steffen, Will, K. Richardson, J. Rockstrom, S. E. Cornell, I. Fetzer, E. M. Bennett, R. Biggs, et al. "Planetary Boundaries: Guiding Human Development on a Changing Planet." *Science* 347, no. 6223 (February 13, 2015): 1259855–1259855. https://doi.org/10.1126/science.1259855.

The Wilderness Act, Pub. L. No. 88–577, § 1131, 1131–1136 U.S.C (1964). https://www.nps.gov/subjects/wilderness/upload/W-Act_508.pdf. Last accessed: April 29, 2024.

Tsing, Anna Lowenhaupt. *The Mushroom at the End of the World: On the Possibility of Life in Capitalist Ruins*. Princeton, NJ: Princeton University Press, 2015.

Wilson, Edward O. *Half-Earth: Our Planet's Fight for Life*. W. W. Norton & Company, 2016.

3

THE HUMAN-TECHNOLOGY EDGE

From Ready-made Artifacts to Dematerialized Humans

Technology emerged as a concept in the early twentieth century and quickly drew attention from many fields of study, including philosophy, history, and sociology. Generally speaking, the understanding of technology has transitioned from internalized models that focused on the essential qualities of technology to externalized models that situate the concepts within relationships in the socio-cultural context. Scholars have attempted to make sense of the role which technology plays in culture and human society.

In the twentieth century, several events shaped society's sentiment toward technology as a whole. During the second Industrial Revolution, roads and railroad networks, telephones, and gas and sewer systems, among many profound technological systems, were widely adopted. Factory electrification enabled modern mass production. These unprecedented human powers fueled a progress narrative underpinning many early works of technology; the narrative presented technological developments as a linear progression, in which one breakthrough led to the next.

Then World War II ended in 1945, with two nuclear bombs dropped upon Japan. For the first time, society experienced the unprecedented destructive power of technology. Meanwhile, the progress narrative proceeded through the following decades; the Apollo 11 moon landing became a global, technological, sublime event through its broadcast in 1969. This ambivalent portrayal of technology led to a 1970s movement known as science, technology, and society (STS). Universities designed new interdisciplinary programs, including the history of technology, women's studies, and the philosophy of technology, to address issues which traditional curricula ignored. The underlying urge was to foster a sense of responsibility in scientists and engineers so that the goals of technological development aligned with the public's best

DOI: 10.4324/9781003320852-5

interests. The implication of the STS movement was more profound than its original context. Toward the late twentieth century, its methods and ways of thinking entered other fields. Many ideas initially developed in STS, such as actor-network theory (ANT), have been appropriated to analyze complex phenomena across disciplines.

In a parallel genealogy, technology became a subject for philosophical investigation when industrialization altered people's life experiences. Philosophers, including Martin Heidegger and Don Ihde, expanded their analyses to technological artifacts and systems in order to understand the role of technology in human existence. Toward the late twentieth century, inspired by AI research and neuroscience, scholars reflected on the meanings of consciousness, intentionality, and mind in the technological lifeworld. For philosophers to study technology is, essentially, to ask what it means to be human. The STS movement and philosophy inspired a range of posthumanist ideas that serve as the theoretical framework for the thesis of the co-production of the environment.

From Mechanic Arts to Systems: A Progress Narrative

With the emergence of *technology* in the English language comes a progressive narrative. The term *technology* was introduced to American vocabulary by Jacob Bigelow in his 1832 *Elements of Technology*. The term then denoted a branch of study concerning mechanical art, as expressed in its Greek roots: *techne* (art or craft) and *ology* (a branch of learning). The term *technology* did not become popular until the 1930s, when an unprecedented complexity in machines occurred. Leo Marx asserts a transition of the meaning of technology from its pre-industrial sense as the study of mechanic arts to its latter-day meaning as the arts themselves. Toward the end of the nineteenth century, applying scientific knowledge to improve the mechanic arts gave rise to an unprecedented form of human power which the term "mechanic arts" failed to convey. Thus, the organizational and material aspects of the semantic void required a concept more inclusive than "machine."[1]

I will analyze Marx's semantic void through two types of complexity in conceptually opposite directions—interior and exterior. Toward the interior, the functional units of a machine become smaller, and through multiple levels of abstraction, the machine becomes a black box that the scientific and technical work becomes invisible. Blackboxing is a term in the Social Construction of Technology (SCOT) to describe the process in which the internal workings of a scientific or technological artifact become opaque and inaccessible when technology and science advance. For example, the smallest functional unit within a computer is an electron flowing through a semiconductor device called a transistor, which works like a switch to control the electrons' passage. This phenomenon can be abstracted into true/false binary values; thus, several transistors can be connected differently to form different logic gates—such as the AND gate or the OR gate. With another level of abstraction, combining various logic gates forms processors that can compute. Without diving into unnecessary details, the point of this example is that when electronic engineers design processors, they do not think about electrons moving through transistors; they care only about logic gates and other higher-level abstractions. When I write computer code, I do not think about logic gates, even though they are the real units performing the functions. Even programming languages nowadays are put into inaccessible black boxes through layers of complexity. A line of code in Python—a popular high-level programming language famous for its human readability—will be interpreted into millions of instructions for the computer, meaning millions of electrons moving through semiconductors. This level of abstraction renders machines like a computer opaque. Ironically, we believe we know how a piece of technology works because we can make it perform certain desired functions, but, at the same time, we do not understand exactly how it works because we cannot see through the levels of abstraction. This opaqueness gives rise to a perceived complexity contributing to the semantic void. If mechanic arts evoke an image of an intricately designed "metal beast" with oily gears and chains, then technology points to a glossy object that performs "scientific witchcraft."

In the exterior direction, we experience a different type of complexity. Electricity must be generated and transmitted to make a computer function. Semiconductor materials—such as silicon—must be mined and manufactured into chips. Other industries design, assemble, and distribute computers. It is challenging, if not impossible, to trace the boundaries of technology. These incomprehensible socio-technical systems constitute additional complexity contributing to the semantic void. We might think we know how a piece of technology works, because it functions in front of our eyes, but at the same time, we cannot trace the full extent of the systems in which the technological artifact is embedded.

This dual complexity gives rise to an unprecedented phenomenon for which we must deploy the term "technology" as a concept to try to capture its essence. Marx argues that the emergence of new technologies is precisely the reason why the concept is hazardous. Because we use this all-encompassing term to represent a conglomerate of items and relations, such generality makes it "peculiarly susceptible to reification."[2] Marx perceives that the danger of the concept of *technology* lies in its deceptive generality, which obscures relationships with humans. Borrowing the idea of "phantom objectivity" from György Lukács, Marx describes the danger of the concept, as if "technologies" possess a certain autonomy that can initiate change, drawing our attention away from who is using these technologies and for what purposes.

Entering the twenty-first century, due to the increasing complexity of technical systems, technology by itself cannot fill the growing semantic void. Once we make this "unprecedented human power" into a thing, the semantic void will not stop growing, and that is the hazardous aspect of technology.[3] By now, we have summoned a collection of concepts into our vocabulary to fill the semantic void: system, network, intelligence, smart, and others on the order of technology. In its broadest sense, systems "are defined by their abstract relations, functions, and information flows, rather than by their concrete material or components."[4] This definition reflects the aforementioned dual complexity. The functionality of systems lies in their relationships and flows rather than in individual material components, such as electrons. If technology conveys a sense of materiality, the term *system* dematerializes this human power into abstracted relationships. Similarly, the term *network* suggests the connectivity of socio-technical systems—a sensing network implies a layer of connected systems embedded in the physical world. Then, how do we know if we are making progress? Smartness becomes a measure. With smart cities becoming a global phenomenon, the Institute of Electrical and Electronics Engineers (IEEE) and the International Organization for Standardization (ISO) have both developed metrics to standardize and thus evaluate the "smartness" of a city.[5] Beyond these global standards, multiple local standards and metrics exist to measure the performance of a city as a socio-technical system. All these concepts specify different aspects of the semantic void which Marx has articulated, forming a roadmap to a technocratic ideology.

Underlying the emergence of technology and its associated terms is a clear progressive narrative. As Marx argues, radical thinkers of the eighteenth century, including Benjamin Franklin and Thomas Jefferson, did not perceive mechanic arts as measures of growth and development. Instead, they viewed machines only as a means to social and political ends. True measures for progress lay in "humanity's step-by-step liberation from aristocratic, ecclesiastical, and monarchic oppression, and the institution of more just, peaceful societies based on the consent of the governed."[6] With the rise of the concept of technology, the distinction between means and ends has gradually dissolved, and technology has become a measure of progress. The transformation formula is again at work, and the concept of technology articulates and rationalizes an unprecedented human power to effectively transform nature. The "phantom objectivity" of technology creates an illusion, as if technology is a thing that empowers those who acquire it. Through this line of reasoning, the end is guaranteed, and all that is left to do is perfect the means—technology.

The progress narrative of technology is a form of technological determinism that asserts that, because of its efficiency, technology essentially determines which aspects of society are best suited for its development.

Technology acts as a factor of "natural selection," culling those aspects that fail to promote it and enhancing those that support its development—this is a form of social Darwinism. A technocratic society may thus progress at the cost of those whose values, morals, and philosophies are less compatible with its development. Technological determinism was further expanded in the mid-twentieth century as a post-war pessimistic view of technoscience after society witnessed the destructive power of technology in modern warfare and environmental degradation.

Technology is the concept that not only distinguishes "advanced" societies from "inferior" ones, but, on a more fundamental level, also distinguishes humans from other species. This distinction establishes a hierarchical ontology that unwittingly justifies a logic that humans need to survive at the expense of "lower" species. The concept of technology and the progress narrative also add to human hubris, in believing that we can end nature. Many argue that industrialization marks the beginning of the Anthropocene, in which humans become the major factor that shapes Earth. In this argument, technology plays an indispensable role. It freezes time, as if the moment we acquired technology, we were on a path of self-destruction. The concept of the Anthropocene is supported by a form of technological determinism, in which progress leads to total collapse. The Anthropocene warns of a self-destructive end in the progress narrative, which humanity must avoid. However, to create the warning, it justifies the progress narratives and transformation formula, reinforces the boundary between human and nonhuman realms, and accepts the "phantom objectivity" of technology.

From Ready-Made Technology to Socio-Technical Systems

Progress narratives posit technology as a transformative force, partly because they treat technologies as "ready-made," thus ignoring the socio-technical ensemble where human agency resides. One sees autonomy in technology and accepts its ability to initiate changes in human relationships and the course of history. The issue of ready-made technology is partly due to how technological development was portrayed in the early twentieth century, when scholars primarily focused on inventions and innovations built around "great man history." The development of scientific discoveries and technological inventions was too often presented as linear progress, with one breakthrough leading to the next.

The founding of the Society of History of Technology (SHOT) in 1958 marked an intellectual milestone. Historians in diverse fields realized that technological innovations had played an essential role in their own narratives, and technology eventually became a subject of historical inquiry in its own right. The field transitioned from an internalized and descriptive model of a "great man history" to an externalized and analytical one. The scholars of SHOT did not find the early-twentieth-century internalist model sufficient for asking fruitful questions; the limited examples of external models could not address the issue, either. For example, Lewis Mumford situates technological innovations within a cultural context in *Technics and Civilization*. He argues that the invention of the clock makes time fungible by structuring and synchronizing human activities, and thus became the basis for capitalism.[7] However, these external models accepted technologies as ready-made, presenting a linear narrative of how technology shaped history; one cannot form more sophisticated questions that do not lead to technological determinism.

The concept of "paradigm shift" altered people's understanding of scientific and technological innovation as a linear progression. Thomas Kuhn, in *The Structure of Scientific Revolutions* (1962), introduces the concept of paradigm: a set of models, worldview, or frameworks agreed upon by a scientific group to define the subject and methods of their study.[8] Doing research in a given paradigm, using Kuhn's term, is doing "normal science," which only furthers the existing theoretical and methodological framework. When different voices challenge the received models, "incommensurability" emerges between competing paradigms. Incommensurability conveys more than differences and disagreements between competing paradigms; instead, it conveys a sense that comparing them is meaningless and impossible due to the semantic difference between them. Though the concepts and terms of the new paradigm are generally inherited from the old one, they take on different meanings and relationships with each other. Communication between the two paradigms becomes challenging, if not impossible, for "the proponents of competing paradigms practice their trades in different worlds."[9] A paradigm shift occurs when the majority of the science community reaches a consensus and agrees on a new set of theories and methods. A paradigm shift asserts that innovations, inventions, and revolutions in science and technology are historically contingent. They consist of intentional choices made by scientific communities, which cannot be separated from their socio-cultural backgrounds.

Kuhn's analysis was itself "a paradigm shift" within the field of the history of technology in the 1970s, when STS emerged as a movement and a field of study in Europe and America. Many scholars took a similar approach, asserting that technological objects were not ready-made, and the direction of technological development was not linear. The STS movement began as a response to a series of events that revealed technology's destructive power in the first half of the twentieth century, including the Dust Bowl in the 1930s and the usage of nuclear weapons during World War II. There had been a robust environmental undertone to the beginning of the STS movement. The first Earth Day was celebrated, the Environmental Protection Agency was founded, and the Clean Air Act was passed in 1970; over the following decades, many further environmental laws and acts were passed. Another acute task for the STS was to critique technological determinism in the mid-twentieth century. The task was clear: to reflect and challenge technological determinism and technocratic ideology by opening the black box of technologies and proving they were socially constructed. The goal was to tie technological artifacts and systems into a socio-technical network, and remind us of the values embedded in the design decisions made by different social groups in developing technologies.

A milestone for this body of research was *The Social Construction of Technological Systems,* edited by Wiebe Bijker, Thomas P. Hughes, and Trevor Pinch as proceedings for a 1984 workshop. Although the scholars in this anthology took on different models to open the black box of technology, they acted under the same banner, known as the Social Construction of Technology (SCOT).[10] Bijker saw the convergence of three different bodies of work. In addition to the STS movement, two others are the UK-based sociology of scientific knowledge (SSK) and the US-based history of technology; all three provide models for understanding technology and society.[11]

For example, Bijker traces the invention of the bicycle.[12] Instead of describing it as a linear process, he analyzes bicycles within several relevant social groups, and each of them interprets the same artifact differently, introducing *interpretative flexibility*. Then each group addressed different problems and goals, providing a way to construct a version of bicycle design—*design flexibility*. This allowed for a wide range of design possibilities, each addressing certain problems (e.g., safety versus speed). In the end, the conflicting designs would reach a conclusion, or a *closure* in SCOT terms, and a standardized design would become the dominant model. A closure occurs either rhetorically (e.g., advertisement) or with the original problems overridden by a new set of problems. After a closure, new social groups would emerge, reintroduce *interpretive flexibility*, and redefine the problems.

We can observe similar phenomena today regarding generative AI systems like Midjourney, DALL-E, and Stable Diffusion. Different approaches exist to build an artificial neural network (ANN) to generate images. Around 2018, Generative Adversarial Network (GAN) was the most popular architecture in the machine-learning community for image generation. In 2018, MIT even listed GAN as one of ten breakthrough technologies.[13] Yet, only four years later, in 2022, diffusion models replaced GAN and became the dominant architecture that undergirds many popular applications. We appear to have reached a closure now when both the machine-learning community and the applications' users agree that diffusion models provide better results.

To account for technological determinism, STS scholars deployed concepts such as *socio-technical ensembles*, *technological culture*, and *technological momentum* to describe the perceived autonomy in technology. Using technological momentum as an example, American technology historian Thomas Hugh proposes that, when a technology is new, it is malleable and subject to social changes; when a mature technology is enmeshed with its society, its deterministic forces make it difficult to influence.[14] Although this view appears to be another way of technological determinism, it does not assert autonomy in technology; instead, technological momentum exists in a socio-technical ensemble. The perceived agency in technology resides in the network of artifacts, standards, regulations, and laws fencing it.

By the end of 2022, the initial excitement about ChatGPT and AI generators was quickly replaced by fear and critique. Artists worldwide participated in a social media boycott against AI art because these AI systems had allegedly been trained with artworks scraped from the internet without their artists' consent. Others feared these AI generators might replace them in the illustration industry. While schoolteachers complained that the students wrote their homework using ChatGPT and argued that this was a form of plagiarism, several academic papers listed ChatGPT as co-authors.[15] Quickly, GPT detectors were introduced to detect AI-generated content, but a study observed that these detectors were biased against non-native English speakers.[16] I serve on the editorial board for an academic journal, and, since early 2023, editors have discussed requiring authors to disclose the use of generative AI systems in their papers. In early 2023, the US Copyright Office granted copyright to "Zarya of the Dawn," a short comic book by Kris Kashtanova with images produced using the AI art generator Midjourney.[17] The Copyright Office granted copyright to the book as a whole but not to the individual images. The Office noted that the AI-generated images on their own lacked human authorship to be copyrighted, whereas the arrangement of these images in creative ways was copyrightable. Kashtanova disagreed and believed that the output of an AI generator was not random and directly depended on the artist's creative input. She said her lawyer would work with her to further explain to the Office "how individual images produced by Midjourney are direct expression(s) of [her] creativity and therefore copyrightable."[18] In 2023, at the beginning stage of generative AI systems, we can already observe a complex socio-technical ensemble around AI systems. Different social groups, as well as policy, law, and culture, act on these generative AI systems and shape their development.

Another SCOT approach is the ANT, developed in the late 1980s by Bruno Latour, Michel Callon, and John Law. Despite the contemporary appropriation of ANT by humanities and social sciences, it was first introduced to the STS community in the same 1984 workshop as one of the models for SCOT. It was only later that Bruno Latour explored the implication of this mode of thinking in a broader context; after that, many other scholars in humanities and social sciences found it useful. Moreover, ANT inspired other popular posthumanist ideas, such as object-oriented ontology (OOO). In Chapter 6, I will discuss ANT in greater detail in the context of nonhuman agency. In SCOT, ANT uses actor-network as a metaphor to describe technology and its role in shaping society. In an ANT framework, both humans and nonhumans are treated equally as "actants." Perhaps unavoidably, the assertion that inanimate nonhumans can "act" has drawn much criticism from the field of STS. However, as I will argue in Chapter 6, most of the criticism is due to limited reading and misinterpretation of ANT, because of its convoluted ontological stance.

The contribution of SCOT is clear. It demonstrates that humans have always coevolved with technology, and that society can steer technological development in a more ethical direction. This body of work duly opened a wave of STS research focused on steering technological development through more active research methods. Since the 2000s, STS has been diversified by different methods and subjects of study; for example, David Guston developed and practiced the idea of "anticipatory governance" in the area of nanotechnology development.[19] Others have bypassed postmodern skepticism about science as a whole and choose to believe the values of the scientific worldview.[20] Harry Collins and others call this "the third wave of science studies," and advocate studying expertise and collaboration among different disciplines. By investigating the concept of *interactional expertise*, they hope to foster a broader collaboration among multiple fields so that technological development may be more transparent and inclusive.[21]

This emerging research is based on a conceptual shift in how scholars study technology and society, from the previous understanding that the two realms are separate entities influencing each other, to a contemporary view that neither can be grasped in isolation—instead, we need to begin with the untidy network of the socio-technical ensemble. It is perhaps more precise to say that scholars have always studied socio-technical systems; from the beginning, the concept of technology has been that of a network of tools and socio-cultural systems, as well as the knowledge and intellect behind them. In a way, the hazardous concept of technology and its "phantom objectivity" are dissolved by STS into a complex but traceable network of systems.

From Technology in Itself to Dematerialized Humans

While STS focuses on returning technology to a socio-technical network, another body of work in philosophy, phenomenology, asks a similar set of questions: What roles do technological artifacts and systems play in everyday human experience?

Martin Heidegger was one of the influential twentieth-century philosophers who addressed technology from a phenomenological perspective. I must define phenomenology in this context to justly comment on Heidegger's approach. In its broadest sense, phenomenology challenges Cartesian subject-object division by positing that "human beings" and "world" are not entities with defined boundaries but are already interwoven and related. Contemporary systems theory may help articulate these philosophical concepts. From a systems perspective, human systems are deeply coupled with other systems, which we call *world* or *environment*. Second-order cybernetics informs us that a human system has no direct access to "the" reality, but merely to a reality constructed by our own system operation; we can only experience "things-for-us," never "things-in-themselves." Thus, in systems theory, systems are structurally open—human beings are always in relationship with the world, but systems are operationally closed so that operating in these relationships determines how the world is disclosed to us. In Heidegger's thesis, "being" is the concept that describes the underlying structure of how humans relate to the world, and how it is disclosed to us. According to Heidegger, "being" is historically determined. "To be" differs in different historical epochs, and this leads to distinct modes of "being" and separate ways in which the world is disclosed to us.

Based on this phenomenological framework, Heidegger traces technology back to its Greek root *techne* and argues that "[t]echnology is a mode of revealing. Technology comes to presence in the realm where revealing and unconcealment take place, where *aletheia*, truth, happens."[22] Again, from a systems perspective, Heidegger's thesis is that technology is one mode for us to construct relationships with other systems and to "disclose" the world. Heidegger believes that modern technology has lost the essence conveyed in *techne*, a way of revealing, or bringing forth, and instead it is about *Gestell* ("enframing"), which entails the sort of technoscience framework or paradigm into which humans are born in the modern technological era.

As one commentator notes, Heidegger did not question "technologies" but "Technology"; Heidegger's analysis was a common approach found in early philosophical thought on technology, when scholars attempted to identify the essential characteristics of technology and its role in culture.[23] There is considerable similarity between Heidegger's approach and the analysis found in the history of technology in the early twentieth century. Lewis Mumford's analysis of time and clock was also a form of reflection on the structure that coordinates human existence. Thus, Heidegger's contribution to technology study should be properly recognized rather than over-interpreted since his understanding did not bypass the paradigm common in the early-twentieth-century study of technology. Like other scholars of his time, Heidegger found autonomy in ready-made technology and could not have foreseen a socio-technical ensemble. He certainly did not anticipate that society could assimilate technology within its operational networks through laws, regulations, and other social functions. At an individual level, Heidegger underestimated human beings' ability to learn and adapt. As Neil Leach rightly puts it:

> [e]ven the most seemingly alienating of technological forms can soon become absorbed within our symbolic horizons, such that they no longer

> appear so alienating...Those who argue that technology is the perpetual source of alienation clearly overlook the potential for human beings to absorb the novel and the unusual within their symbolic framework.[24]

Fundamentally, Heidegger's comments on technology assumed an unspoiled origin when human beings existed in a "naked body" with a "naked mind." He ignored the realities that humans and human societies have always co-evolved with tools, structures, languages, systems, and technologies; this condition defines who we are as humans. It may be too extreme to "forget Heidegger" as Neil Leach demands, but we do need to properly historicize Heidegger's "questions concerning technology" before developing a novel framework to understand what it means to be a contemporary human being.

In fact, unlike Heidegger, who was nostalgic about the original human condition and critical of modern technology, later philosophers saw potential in modern technology to consider what being human means. Philosopher Don Ihde studies technology within a phenomenological framework but does not trace it to the past as Heidegger did. Instead, Ihde looks forward, and enquires about the new relationships which technological artifacts and systems open, so that we can "reveal" the world differently with modern technology. In a "human-world" phenomenological framework, Ihde has developed three categories of phenomenological relations between humans and technology: mediated relations, alterity relations, and background relations. Furthermore, within mediated relations, two subcategories exist: embodiment relation and hermeneutic relation.[25] These relations are illustrated in Figure 3.1.

In embodiment relation, technology broadens or dampens the sensitivity of the human body to the world. In Ihde's example, "I wearing glasses," the

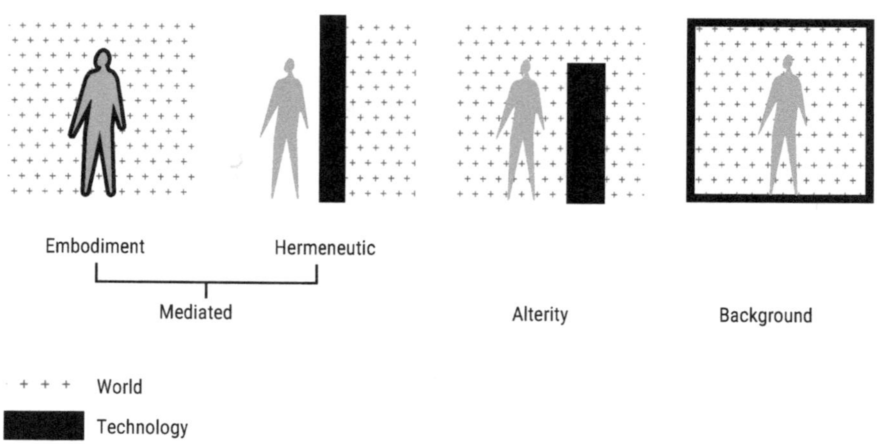

FIGURE 3.1 Don Ihde's phenomenology of technology. Illustration by the author.

human does not look *at* glasses but looks (*through* them) at the world. Ihde formalized this relation as:

(I-technology) -> world

In the hermeneutic relation, technology provides a representation of certain aspects of the world. Ihde argues that a thermometer reveals temperature, and when we peer at a thermometer, we do not look at it, but instead at one aspect of the world—the ambient temperature. "Hermeneutic" signifies that this relation involves interpretation. We must first mentally construct what 25°C means, then experience the world through a thermometer as a medium. Ihde formalizes hermeneutic relation as:

I -> (technology-world)

In alterity relations humans do not relate to the world via the medium of technology. Instead, we relate to technologies as if they are "others." Alterity relations are formalized as:

I -> technology (-world)

In this relation, Ihde articulates a sense of anthropomorphism; we tend to project human properties and emotions onto technological artifacts. Technologies behave as "quasi-other" because they provide a degree of interaction, and yet assume a certain degree of independence.[26] Ihde's argument is akin to a concept in robotics called "perceived intelligence," which describes how intelligent a robot appears. To robotics engineers, who know the rules and algorithms that drive a robot, the robot does not appear intelligent at all, whereas to a lay audience, the robot appears intelligent enough to relate to as an independent other. Today, generative AI systems like ChatGPT pose more challenging questions about alterity relations. In several cases, even computer scientists do not know why a machine-learning model behaves in specific ways, and alterity relations thus become salient.

The last type of relations is background relations, in which technologies play no central role in our experience of the world. Instead, they recede into the background and shape the context of our experience. For example, a heating system maintains the environment in a stable condition, so it backgrounds our experience. In this relation, we do not even notice technologies except where they malfunction. Ihde formalizes background as follows:

I- (technology/world)[27]

The background relations touch on one aspect of the cybernetic environment, where machines become active players in constructing the shared environment, thus providing a background for human experience.

Ihde's scheme of four relations should not be applied in isolation. We may experience more than one type of relationship with a single technological artifact. A smartphone perfectly illustrates how the above examples mesh into a single object. We check outside temperatures via a smartphone (hermeneutic relations). Taking photos, we regard the world through the phone's camera lens (embodiment relations). A smartphone is intelligent enough to be treated as an independent other, especially when digital companions such as Google Assistant or Siri are enabled (alterity relations). Finally, we carry our smartphones, which constantly ping GPS signals. These GPS coordinates are utilized by programs such as Google Maps to determine traffic conditions and calculate the best route for driving. This "traffic map" is a constructed reality that serves as a background context for navigating the world. In our daily lived experience, smartphones thus establish essential nodes of a technological system of way-finding (background relations). In this example, Ihde's scheme is a useful framework for comprehending our daily experiences, but it is only a starting point in evoking a more sophisticated understanding of human-technology relations. With technological advances such as sensing networks, AI, machine learning, and cyberphysical systems, reflecting on our experiences using Ihde's scheme alone is cumbersome.

Another concept—the extended mind—may not at first glance appear to be related to the concept of technology, but it is worth exploring. Informed by advances in cognitive science, philosophers Andy Clark and David Chalmers propose that the environment functions as part of the human mind.[28] They argue that a person with Alzheimer's disease, who relies on a notebook to "remember," is, in terms of cognitive processes, no different from an individual with a good memory. In the case of the person with Alzheimer's, their internal mental processes and notebook may be regarded as a single system. Even though information within the notebook, in this case, must be accessed by perception (eyesight), the perceptual processes may be considered as internal information flows of the human-notebook system. Because systems are defined by relations rather than material components, from a cognitive system point of view, it matters little whether the system consists of a "naked mind" or a human-notebook assemblage. Neither needs anything outside the cognitive system to perform its functions. From this perspective, Clark and Chalmers argue that humans can offload cognitive functions onto the environment. It is not humans who are intelligent, but human-environment assemblages make us appear intelligent.[29]

One classic example along this line of thinking is the thought experiment of the "Chinese room." A non-Chinese speaker fluent in English is locked in a room with a Chinese-English dictionary and a hypothetical rulebook of Chinese-English translation. A note bearing Chinese characters is slipped under the door, and the entrapped person must follow the instructions and rules in order to translate the characters. John Searle, the original author of this thought experiment, tried to argue that a person who does not speak Chinese at all can easily pass a Turing test and "fool" those outside the room into believing they understand Chinese. Similar conditions may apply to AI that translates Chinese; yet Searle wanted to prove that this AI cannot understand the Chinese language and thus neither "thinks" nor possesses a mind. However, from the extended mind perspective, it is not that a human understands Chinese but that the human-book-room assemblage does so. From a systems perspective, the "human" does not matter. On the contrary, it is quite a "human" thing to rely on the environment to appear more intelligent.

Rather than reflections on technology, philosophical investigations are, more precisely, reflections on humans. The two examples—extended mind and human-technology relations—demonstrate that if a boundary exists between technology and humans, it must be more fluid than we thought. Since humans are constantly involved in a network of socio-technical systems, any concept that maintains the notions of a "naked mind" and "naked body" prevents us from understanding the human condition in the technological lifeworld. We have always been cyborgs.

Notes

1 Marx, "Technology: The Emergence of a Hazardous Concept."
2 Marx, 576.
3 Marx, "Technology: The Emergence of a Hazardous Concept."
4 Heylighen and Joslyn, "Cybernetics and Second-Order Cybernetics," 5.
5 See, for example, https://smartcities.ieee.org/. Last accessed: April 29, 2024.
6 Marx, "Technology: The Emergence of a Hazardous Concept," 565.
7 Mumford, *Technics and Civilization.*
8 Kuhn, *The Structure of Scientific Revolutions: 50th Anniversary Edition.*
9 Kuhn, 149.
10 Bijker, Hughes, and Pinch, *The Social Construction of Technological Systems.*
11 Bijker, "How Is Technology Made?"
12 Bijker, Of Bicycles, Bakelites, and Bulbs.
13 "10 Breakthrough Technologies 2018." https://www.technologyreview.com/10-breakthrough-technologies/2018/. Last accessed: April 29, 2024.
14 Bijker, "How Is Technology Made?"
15 Generative Pre-trained Transformer and Zhavoronkov, "Rapamycin in the Context of Pascal's Wager"; Kung et al., "Performance of ChatGPT on USMLE"; O'Connor and ChatGPT, "Open Artificial Intelligence Platforms in Nursing Education."
16 Liang et al., "GPT Detectors Are Biased against Non-Native English Writers."
17 https://www.kris.art/portfolio-2/project-one-ephnc-jamy8. Last accessed: April 29, 2024.
18 https://www.instagram.com/p/Co-aYkQumio/?utm_source=ig_embed&ig_rid=22750dbf-0de5-4f1c-8b49-b189d1e7c793. Last accessed: April 29, 2024.
19 Guston, "Understanding 'Anticipatory Governance.'"
20 Collins, "We Cannot Live by Scepticism Alone."
21 Collins and Evans, "The Third Wave of Science Studies"; Collins, Evans, and Gorman, "Trading Zones and Interactional Expertise"; Gorman, *Trading Zones and Interactional Expertise: Creating New Kinds of Collaboration.*
22 Heidegger, "The Question Concerning Technology," 295.

23 Verbeek, "Don Ihde: The Technological Lifeworld."
24 Leach, "Forget Heidegger."
25 Ihde, *Technology and the Lifeworld: From Garden to Earth.*
26 Ihde.
27 Ihde.
28 Clark and Chalmers, "The Extended Mind."
29 Clark and Chalmers.

Bibliography

Bijker, Wiebe E. "How Is Technology Made?—That Is the Question!" *Cambridge Journal of Economics* 34, no. 1 (January 1, 2010): 63–76. https://doi.org/10.1093/cje/bep068.

———. *Of Bicycles, Bakelites, and Bulbs: Toward a Theory of Sociotechnical Change.* First Paperback Edition. Cambridge, MA: The MIT Press, 1997.

———. Thomas Parke Hughes, and Trevor Pinch, eds. *The Social Construction of Technological Systems: New Directions in the Sociology and History of Technology.* Cambridge, MA: MIT Press, 1987.

Clark, Andy, and David Chalmers. "The Extended Mind." *Analysis* 58, no. 1 (January 1, 1998): 7–19. https://doi.org/10.1093/analys/58.1.7.

Collins, Harry. "We Cannot Live by Scepticism Alone." Comments and Opinion. *Nature* (March 4, 2009). https://doi.org/10.1038/458030a.

Collins, Harry, and Robert Evans. "The Third Wave of Science Studies: Studies of Expertise and Experience." *Social Studies of Science* 32, no. 2 (April 1, 2002): 235–96. https://doi.org/10.1177/0306312702032002003.

Collins, Harry, Robert Evans, and Michael E Gorman. "Trading Zones and Interactional Expertise." In *Trading Zones and Interactional Expertise: Creating New Kinds of Collaboration*, edited by Michael E Gorman. Inside Technology. Cambridge, MA: MIT Press, 2010.

Generative Pre-trained Transformer, ChatGPT, and Alex Zhavoronkov. "Rapamycin in the Context of Pascal's Wager: Generative Pre-Trained Transformer Perspective." *Oncoscience* 9 (December 21, 2022): 82–84. https://doi.org/10.18632/oncoscience.571.

Gorman, Michael E. *Trading Zones and Interactional Expertise: Creating New Kinds of Collaboration. Inside Technology.* Cambridge, MA: MIT Press, 2010.

Guston, David H. "Understanding 'Anticipatory Governance.'" *Social Studies of Science* 44, no. 2 (April 2014): 218–42. https://doi.org/10.1177/0306312713508669.

Heidegger, Martin. "The Question Concerning Technology." In *Martin Heidegger Basic Writings*, edited by David Farrell Krell, translated by Walter Lovitt, 307–41. New York: HarperCollins, 1977.

Heylighen, Francis, and Cliff Joslyn. "Cybernetics and Second-Order Cybernetics." In *Encyclopedia of Physical Science and Technology*, 155–69. Amsterdam: Elsevier, 2003. https://doi.org/10.1016/B0-12-227410-5/00161-7.

Ihde, Don. *Technology and the Lifeworld: From Garden to Earth.* Midland Book. Bloomington: Indiana University Press, 1990.

Kuhn, Thomas S. *The Structure of Scientific Revolutions: 50th Anniversary Edition.* Chicago, IL: University of Chicago Press, 2012.

Kung, Tiffany H., Morgan Cheatham, ChatGPT, Arielle Medenilla, Czarina Sillos, Lorie De Leon, Camille Elepaño, et al. "Performance of ChatGPT on USMLE:

Potential for AI-Assisted Medical Education Using Large Language Models." medRxiv, December 21, 2022. https://doi.org/10.1101/2022.12.19.22283643.
Leach, Neil. "Forget Heidegger." In *Designing for a Digital World*, edited by Neil Leach. London: Wiley-Academy, 2002.
Liang, Weixin, Mert Yuksekgonul, Yining Mao, Eric Wu, and James Zou. "GPT Detectors Are Biased against Non-Native English Writers." *Patterns* 0, no. 0 (July 10, 2023). https://doi.org/10.1016/j.patter.2023.100779.
Marx, Leo. "Technology: The Emergence of a Hazardous Concept." *Technology and Culture* 51, no. 3 (2010): 561–77. https://doi.org/10.1353/tech.2010.0009.
MIT Technology Review. "10 Breakthrough Technologies 2018." Accessed July 24, 2023. https://www.technologyreview.com/10-breakthrough-technologies/2018/.
Mumford, Lewis. *Technics and Civilization*. New York: Harcourt, Brace and company, 1934.
O'Connor, Siobhan, and ChatGPT. "Open Artificial Intelligence Platforms in Nursing Education: Tools for Academic Progress or Abuse?" *Nurse Education in Practice* 66 (January 1, 2023): 103537. https://doi.org/10.1016/j.nepr.2022.103537.
Verbeek, Peter-Paul. "Don Ihde: The Technological Lifeworld." In *American Philosophy of Technology: The Empirical Turn*, edited by Hans Achterhuis, 119–46. The Indiana Series in the Philosophy of Technology. Bloomington: Indiana University Press, 2001.

4

THE HUMAN-NATURE EDGE

The Three Waves of "Nature Study"

In this chapter, I will trace the evolution of the ideas about nature in the twentieth and early twenty-first centuries along the edge of human-nature to understand the role of "nature" in present-day environmental narratives. I characterize this development of ideas as three waves of reflection, represented in Figure 4.1.

The first wave was a transition within the ecological sciences in the 1950s, when a non-equilibrium model replaced the earlier homeostatic view of nature, exemplified by Gleason's individualistic model replacing Clements's succession model. The second wave was a constructivist reflection on "nature" in the 1990s by environmental historians and STS scholars, who claim "nature" is socially constructed by exposing unstated social and cultural values embedded in that term during the environmentalism movement. The third wave reflects on the radical claim of the social construction of nature, which gave rise to the early twenty-first century "nonhuman turn" and a series of posthumanist epistemology and ontology arguments. Contemporary posthumanist intellects work within various frameworks: political ecologies, multispecies ethnography, animal studies, feminist materialism, and object-oriented philosophy. In these posthumanist frameworks, the human becomes a product of the co-evolution, physically and conceptually, with other nonhuman species. Nature may be a constructed concept, but the nonhuman realm is also real; what we have understood as "nature" has forever been a co-evolving, co-producing, and interweaving network of humanity and non-humanity.

Each wave represents a development of the concepts of nature and its relationships with humans. The frontier of each wave mobilizes across different fields of studies. By reflecting on the previous wave of transition, scholars in

DOI: 10.4324/9781003320852-6

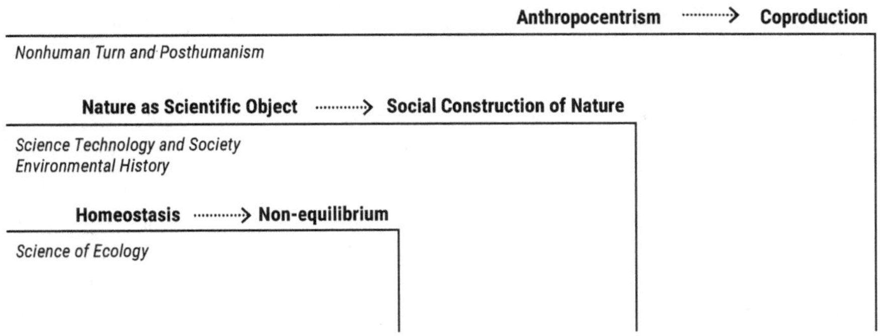

FIGURE 4.1 Three waves of conceptualizations of nature. Illustration by the author.

the new wave build a higher order of reflexivity about human-nature relationships. The ideas generated in the new wave do not directly replace the old ones; instead, they provide a different perspective of "nature" to the already diverse field. In this maturing process, with perspectives added to the discourse, the concept of nature grows complex and, at the same time, contested. Together, the three waves of transition have contributed models that constitute our contemporary understanding of human-nature relations.

Before we proceed, I must emphasize the strong environmental undertone since the mid-twentieth century that has provided strong moral imperatives for building a more sophisticated understanding of nature. Stories of DDT and extinct species, Rachel Carson's *Silent Spring*, the first Earth Day on April 22, 1970, and the establishment of the U.S. Environmental Protection Agency (EPA) in December 1970, prepared the American psyche for a profound environmental movement. Early concerns regarding air and water pollution evolved into concerns over climate change. In response came environmental policies and concepts such as the Clean Air Act (1963) and Clean Water Act (1972); sustainable development in the 1990s; stormwater management and low-impact development (LID) toward the early 2000s; and, dating from the late aughts, resilience, and adaptivity. These terms and concepts constitute today's environmental discourse, constructing versions of "nature." One message I want to convey in this chapter is that these concepts, including certain popular ones in contemporary environmental discourse, such as sustainability, are not free from critique. They reinforce the transformation formula: not only do humans transform nature, but we should also sustain and protect it from further transformation.

Another caveat I must comment upon is landscape design as a cultural practice that reflects society's understanding of nature. We may compare the three waves of conceptual development with the paradigm shifts in the landscape discipline. Landscape architecture took on an ecological model in the mid-twentieth century, thanks to Ian McHarg's teaching and practice. Designers began to explore how time and succession work to their advantage. When a non-equilibrium view gained currency in ecological science, landscape designers began to reevaluate earlier ecological determinism and explored more dynamic and open-ended frameworks. Following the second wave of transition, landscape architects reflected on the dark side of environmentalism, such as uncritical conservation, preservation, and restoration efforts. Many began to construct experiences, rather than natural scenery, and expanded aesthetic categories that connected people to the biophysical world surrounding them. In response to the third wave of "nonhuman turn," multispecies interactions and coproduction of landscapes have become a conceptual framework for a new generation of landscape architects.

However, these transitions in the broader discourse of nature and landscape architecture discipline must not be viewed as one paradigm replacing the other. Environmental historians' reflections do not fundamentally alter how ecologists conduct scientific research, and posthumanist ideas do not prevent constructivists from uncovering alternative histories of a culturally constructed nature. Similarly, in landscape architecture, different ways of thinking and modes of practice inspire designers to design landscapes that reflect their understanding of "nature." Although each wave of transition abandons concepts, the three shifting waves contribute to a more comprehensive view of human-nature relations. The transformation formula, to a greater extent, becomes an amalgamation of outdated views on nature.

The First Wave: From Homeostasis to Open-Endedness

For the first half of the twentieth century, nature comprised the exterior, nonhuman realm, with humans disturbing it. Toward the late twentieth century, many recognized human activity as an important part of the complex system called nature. These transitions occurred mainly in the science of ecology, as well as in ecology-based environmental practices. I will analyze both fields to illustrate the first wave of transition.

During most of the twentieth century, the nonhuman realm was investigated below scientific apparatuses, and humans were observing subjects on the other side. Within the subject-object framework, "nature" became an all-encompassing concept that included the nonhuman realm merely as a scientific object. Within this scientific tradition, the emergence of ecological thinking gave rise to the scientific discipline of ecology in the early twentieth century. Ecological thinking emerged from the convergence of natural history and natural philosophy, or science, in the nineteenth century. Naturalists such as Alexander Humboldt (1769–1859) introduced Newtonian analytic reasoning and measurement to natural history, and informed a new paradigm of scientific knowledge production. The study of nature transitioned from descriptions of nature through taxonomy, to a more analytical form of study that attempted to use data to explain the underlying "law of nature."[1] In the early twentieth century, ecology grew into an independent scientific discipline, providing a model of thinking that asked a unique set of questions, and produced a new category of knowledge about the nonhuman realm. Instead of describing what an object was made of, such as developing the anatomy of a plant by naming its parts, ecology was concerned with how the object functioned in an extensive system or network, and, in turn, how the system and network influenced its behavior.

Early ecological research focused primarily on plant ecology and plant succession. In the early twentieth century, two competing models for ecological succession existed: a holistic or organismic model proposed by Frederic Clements (1874–1945), and an individualistic view held by Henry Gleason (1882–1975). Clements's model used "organism" as a metaphor, asserting that species in an ecosystem were deeply connected, and the ecosystem as a whole functioned as an organism. After disturbances such as wildfires, the ecosystem would repeat a series of stages and reach an equilibrium or a climax formation. In contrast, Gleason believed that plant species required different environmental conditions, and ecosystem dynamics resulted from individual species responding to environmental factors. Therefore, co-habiting plants did not form communities, and observed associations were simply coincidences. Gleason's individualistic model was largely ignored until the early 1950s, when ecologist John Curtis further developed Gleason's concept. Now, many have espoused a more dynamic view of physical landscapes, viewing them as continually shifting mosaics, with different ecosystem types emerging and disappearing.[2]

Though Clements's hypothesis was not entirely wrong, Gleason's individualistic concepts still superseded Clements's organismic concept; Clements's ideas, such as "climax community," have been largely abandoned. As reported by many ecologists, they observe patterns and collect data to support both views in their field practice. Several have even reported that Clements's model was, in fact, more effective in restoration and preservation ecology.[3] Further evidence appears to suggest that plants are more connected than previously thought. For example, since the 1980s, mycorrhizal networks have become a trending topic of research in ecology. Mycorrhizal fungi create underground hyphal networks that connect individual plants, and plants use this network to "communicate" with each other, transferring water, nutrients, minerals, allelochemicals, and defense signals.[4] Mycorrhizal networks demonstrate that individual species do connect to each other, and on a level more literal than Clements's sense.

In a way, both Clements and Gleason told portions of the story. Many ecologists now agree that plants in different places, with different spatial and temporal scales, can exhibit different degrees of connectedness as well as interdependency; thus, Gleason and Clements can be placed at different positions on a spectrum, between randomness and structure.[5] If we observe across different scales, temporal and spatial, what we see as stable is merely a relatively stable state in the course of evolution. For example, researchers have framed future sea-level rise as a threat to coastal cities, and scientists project that, by the end of this century, the global average sea level will have risen approximately one meter. However, also according to researchers, the sea level has climbed more than 120 m over the past 20,000 years. It was not until 5,000 years ago that sea levels reached a relatively stable state, when humans began to build concentrated settlements.[6] We have studied coastal ecology for fewer than 200 years. Hence, on a geological time scale, what we have believed is a stable state is merely an observation confined by the thresholds of human perception.

If Gleason's model did not completely replace Clements's concept, what has changed in this paradigm shift? I must analyze their underlying conceptions of nature to help answer that question. Clements's organism metaphor was based on a conception that nature is a static entity, and its "pristine power" returned it to post-disturbance equilibrium. Clements preferred order and predictability. In contrast, Gleason's model emphasized chance and embraced indeterminacy; it portrayed a version of nature that was less predictable and in constant flux. In this paradigm shift, what fundamentally changed was our tolerance for unpredictability. In other words, it was not that Gleason's model was closer to the truth, but after the 1950s, its underlying conceptions of nature found favor with other ecologists. "A non-equilibrium view of natural processes has literally changed the way scientists think about the nature of nature; they now frequently see change as probabilistic and multidirectional, rather than as a progressive march toward clear endpoints."[7]

Another aspect of the first wave was that the position of humans in nature had changed. Ecology as a scientific discipline bore an early subject-object structural influence within science, and, until recent years, ecologists have studied a version of nature without humans in it. Over the past three decades, ecologists have begun to pay attention to human activities as essential parts of the ecological process. For example, environmental scientist Lance Gunderson and ecologist C. S. Holling coined *panarchy* to explain complex relationships between human and natural systems. Panarchy articulates a series of continual adaptive cycles of growth, accumulation, restructuring, and renewal, in interlinked human-nature systems.[8] It is a concept that underpins many contemporary terms in environmental discourse, such as adaptation, resilience, and adaptive management. Similarly, "anthroecology," a term proposed by environmental scientist Erle Ellis, advocates studying ecology under the premise that humans have always been important to ecosystems.

Anthroecology reflects the epistemology of ecological science, and challenges ecologists to confront the muddle of the anthroecosystem.[9]

Admittedly, these scientists might have deployed better terms than "human-nature system" and "anthroecology," because they seemingly suggest combining human and nature, when, in fact, the scientists intended to challenge people to talk about nature with humans as an important ingredient.

We also observe the shift of concepts of nature in practice-oriented disciplines, such as landscape architecture. The notion of "pristine nature" provided moral imperatives for a plethora of environmental and ecological practices. Perceived stability in the ecosystem was often rendered useful and provided a baseline to measure the imagined pristine state; thus, many environmental practices tend to encourage this apparent stable state through preservation, conservation, and restoration efforts. Since the mid-1950s, the discipline of landscape architecture has adopted ecology as a model with which to approach landscape design. Landscape architect Ian McHarg introduced ecology and scientific methods into landscape architecture through teaching and practice in the mid-twentieth century. McHarg updated the map overlay method with ecological thinking, to form an organizing framework to examine the interactions of nature's "layers," including geology, hydrology, vegetation, soil, and land use. The McHargian design method was essentially a reworking of Clements's ecological succession model, given that Ian McHarg began practicing and lecturing in the 1950s, when Clements's model still dominated the understanding of ecological succession. It is a version of ecological determinism that implies a linear progression of ecosystem evolution; given the right environmental conditions, ecosystems may always reach the desired primal state. Using solid boundaries and shadings to represent different features unavoidably freezes space and time and presents a static view of nature.

In his seminal book, *Design with Nature*, McHarg repeatedly utilizes the concept of the climax stage to describe ecological succession with a definite end.[10] Moreover, throughout his book, he positions humans and nature against each other, describing them as "diametrically different environments, the poles of man [*sic*] and nature."[11] McHarg used compelling rhetoric to advocate that landscape architects help "nature" achieve this desired homeostatic state through design choices. The McHargian design approach represents many ecological practices in line with the transformation formula; by maintaining one of many ecosystem states, humans can further benefit from perceived stability.

The next generations of landscape architects inherited the McHargian ecological design framework, but updated their ecological concepts and fundamental conceptions of nature. First, many have adopted an emergent view of ecosystem dynamics. Rather than projecting a clear endpoint and design for a stable state, these designers recognize that ecosystems are in constant flux and evolve over time, so design strategies should promote this evolution and account for emergence. For example, in their entry for the Toronto Downsview competition (1999), the Field Operation and Stan Allen, working with ecologist Nina-Marie Lister, proposed a framework that reflects this shift of conception. The proposal "imagined physical scaffolds that would sponsor the propagation of emergent ecologies, natural systems that would be seeded initially and evolve over time with increasing levels of complexity and adaptability."[12] Landscape architects are no longer interested in designing a stable ecosystem as McHarg did; instead, they prefer to design a flexible framework that allows ecosystems to evolve. Second, many have recognized human activities and participation in system evolution. Field Operation's emergent framework is achieved through an adaptive management strategy, which requires constant human maintenance based on long-term monitoring and observation. In contrast to the McHargian design paradigm, which relies on "the power of pristine nature" to undo human disturbances, the new paradigm of ecology-based design envisages frameworks that allow humans to actively engage in ecological processes.

In summary, in the first wave of the shift, the conception that "nature is a static entity out there with humans disturbing (transforming) it" was abandoned. Regarding the transformation formula, the "pristine" quality is merely a perceived stable state on a specific spatial and temporal scale. Humans do not transform nature; we are always among the forces that give rise to the "pristine."

The Second Wave: From Scientific Fact to the Social Construction of Nature

A critical transition in the second wave of the shift is the ascendance of constructivist epistemology and the understanding of nature as a social construct. Second-order cybernetics and the STS movement led to a broad-based reflection on knowledge and scientific facts, claiming that society actively constructs knowledge. Scientists do not reveal facts. Instead, these facts are consensus-reached within a scientific community.

The concept of paradigm may further illustrate how societal values influence scientific research. Chapter 3 introduced Thomas Kuhn's paradigm; here, I provide a brief summary. Introduced by Kuhn in *The Structure of Scientific Revolutions*, a paradigm is a set of models, worldviews, or frameworks agreed upon by a scientific group, to define the subject and the method of their study.[13] Doing research in a given paradigm is doing "normal science," which further articulates the existing theoretical and methodological framework. A paradigm shift occurs when the majority of the scientific community reaches a consensus on a new set of theories and methods. Paradigm shift implies that revolutions in scientific research are historically contingent on choices made by scientific communities, which are inseparable from their socio-cultural backgrounds, values, and belief systems.

The ecological paradigm shift in the 1950s mirrored a broader socio-cultural change in post-World War II America. American society began to embrace "technological, social, and cultural actions that celebrated individualism, rebellion from previous norms, and a profound acceptance of uncertainty."[14] This societal movement manifested in many aspects, including the civil rights movement that embraced individual importance; anti-socialism; laissez-faire individualism on the political stage; expressionism and existentialism in art and philosophy; and the rise of post-structuralism in the intellectual arena.[15] Moreover, since the mid-twentieth century, developments in research, such as second-order cybernetics, general systems theory, and chaos theory, all contributed to the American psyche to embrace individualism and revolt against conventions and norms. Scholars were ready to alter their explanations of how nature works, from a perspective that promotes individualism and indeterminism.

Ecologists' understandings of how the nonhuman realm works depend on the lens they deploy to observe it, but the lens they choose is essentially an image of the society, laden with socially and culturally specific values. The paradigm shift in the science of ecology serves as evidence for the thesis of the "social construction of nature."

Taking on the constructivist mentality reveals that "nature" is problematic. When we speak of the nature of something, we imply an end of a dispute, a final definition, and an ideal form of how something ought to be. Then, when we pour culturally specific values into the term "nature," we assume these values to be universally true.[16] It is these values that underpin how models of nature are constructed and employed to justify our actions. One cornerstone in the second wave of reflection is the anthology *Uncommon Ground: Rethinking the Human Place in Nature*. The entries posed critical questions: Whose nature to protect? To which time of history do we recover? What values have we poured into the term nature? "Wilderness" was one of the themes challenged in the anthology. Many conservation and preservation narratives were underpinned by a myth of "pristine nature" that was worth protection, and thus wilderness played an irreplaceable role in environmental narratives. However, when environmentalists regarded the Amazon rainforest as the "jungle" and wilderness, they were, in fact, speaking about the homes of indigenous people who had transformed and managed the rainforest for centuries.[17] Beyond wilderness, the entries in *Uncommon Ground* also explored topics such as the Edenic theme in Western culture, the commodification of nature, simulated nature, and other critical and occasionally controversial issues.

The role of landscape architecture in constructing nature was also recognized and accurately reflected by one contributor, landscape scholar Anne

Whiston Spirn. Spirn examined Frederic Law Olmsted's practice in the late nineteenth and early twentieth centuries, and recognized the profession's contribution to translating cultural understandings of nature into public landscapes. More importantly, she revealed Olmsted's double-edged legacy. On the one hand, Olmsted established a tradition in the landscape profession of creatively applying environmental knowledge. On the other hand, his creativity was based on an image of the landscape where the design spaces resembled natural scenery, and thus were not adequately valued as human constructs. To use Spirn's example, people refuse to cut down the trees Olmsted planned to cull in his original design, because they assumed the built landscape was natural.[18] By constructing "natural scenery," Olmsted unwittingly reinforced an image of pristine nature that belonged to a small fraction of humans with the privilege to conceptualize. Here, the landscape image became a repository for the socio-cultural values of nature.

Spirn's analysis reflects critical questions posed to the landscape architecture profession during the 1990s. Does landscape design simply reinforce a biased version of nature? If not, then what should be the role of this discipline? In response to these questions, a new narrative has emerged to articulate the role of landscape architects in contemporary society. The broader goal for landscape architects is to expand the category of aesthetic experiences of nature beyond the picturesque, sublime, and beautiful, and to cultivate new definitions of aesthetics that promote an ecological and biocentric worldview.[19]

Teardrop Park in New York City, opened in 2004 and designed by Michael Van Valkenburg Associates, is an example of heightened natural experiences. The 27-foot-high, 168-foot-long stacked bluestone Ice-Water Wall creates a displacement of local geology and display of microclimate. Because this retaining wall faces northeast, cold weather in wintry New York City freezes the water seeping through the bluestone stacks. Ice forms on the wall, resonating with the phenomena found in the mountains of upstate New York. Here, the design does not reconstruct natural scenery—layered rock piles and stacked bluestone walls resemble nothing natural. Instead, the wall elucidates and amplifies freeze-thaw phenomena, using constructed landforms and microclimates, and the designed landscape encourages people to find excitement in the mundane natural phenomena surrounding them. This design may be juxtaposed with William Cronon's arguments: "[t]o think ourselves capable of causing 'the end of nature' is an act of great hubris, for it means forgetting the wildness that dwells everywhere within and around us."[20] To prevent "the end of nature" serves as a moral imperative to protect the wilderness, but protecting a "wilderness" forces people to ignore the "wildness" around them. Teardrop Park's ice wall pushes the boundary of what it means to construct nature beyond beautiful scenery, by directing people's attention to the "wildness" within an urban environment.

While environmental historians provided verbal critiques, contemporary landscape projects such as Teardrop Park serve as physical responses to the darker side of environmentalism. In the constructivist narrative, landscapes are by-products of culture and repositories of values. New generations of landscape architects provide an alternative and more positive way to understand the social construction of nature. Projects like Teardrop Park are evidence that society can construct versions of nature that are not pristine and recognize the wildness of the nonhuman realm through designed experiences. Landscapes should be not only repositories of cultural values but also experiences that are didactic, teaching new morals, ethics, and ecological values.

In the second wave of the shift, social constructivism ascended as a conceptual framework to elucidate nature and its cultural meaning. In the transformation formula, "pristine nature" is a culturally specific concept constructed by a fraction of humanity. For others, "pristine nature" might be their home, and transformation means exploitation. Because of the environmental movement, "pristine nature" becomes a myth in which to inject all sorts of environmental values. When people transfer these values into actions, they reinforce a version of nature that has already been narrowly defined.

In North America, pristine nature points to a time when the colonial settlers discovered the "new" world. Concepts such as wilderness were constructed to reinforce this origin, and many environmental practices are justified to protect it. The social construction of nature reveals an obstruction difficult to overcome, because to challenge the transformation formula in the American context is to challenge an origin myth. However, what we perceived as pristine was produced by those who came before us, human or nonhuman; what we thought natural was already invested with human values. Most importantly, if we focus too heavily on "pristine nature" in a distant place, we ignore the wild nonhumans around us and how they influence the shared environment in meaningful ways.

The Third Wave: From Anthropocentrism to Nonhuman Agency

One incentive for radical constructivism is to reflect on human hubris, and place a limit on human knowing. However, late-twentieth-century constructivism backfired, because to claim that nature is a social construct asserts, ironically, another level of human hubris. Here is the irony: using the term "nature" to study the nonhuman realm, as the constructivists also realized, is eventually to study a set of human values. No matter how reflective our efforts, the nonhuman realm remains outside our conceptual framework of human language and thinking. The authors whose work appears in *Uncommon Ground* too, recognized this irony and the limitations of their constructivist framework. For example, when Cronon encouraged readers to focus on the "wildness" within and around them, he attempted to devise another framework to study nonhumans—a framework focused on the intrinsic values of nonhumans, free from human standards. As Spirn noted, "[t]here is always a tension in landscape between the reality and autonomy of the nonhuman and its cultural construction... nature may be constructed, but it is not *only* a construction."[21]

The way scholars approach the nonhuman realm in the twentieth century has been influenced by a deeper structure—the humanist worldview since the Enlightenment. The hierarchical ontology and anthropocentric values justify human survival at the expense of those on "lower" levels of the survival pyramid. The humanist view results in a deep-rooted irony that haunts the contemporary environmental discourse; the nonhuman realm remains a necessary resource for human use, and environmental practices merely provide patches that sustain an outdated socio-cultural system, be it capitalism or neoliberalism, to keep up with exploitation of other species and their habitats. The view of nature as a resource subject to human use reinforces the transformation formula by associating profit with transformation. The *usership* narrative is diversified by a *stewardship* narrative: not only should we use nature as resources, but sustain it for further use. This narrative deceptively conveys both a sense of right and one of obligation to protect "nature" for further exploitation. Ecological thinking and practices are reduced to indispensable adjuncts of exploitation, rather than a lens through which we may recognize the autonomy and agency of the nonhuman realm.

For example, the concept of "ecosystem services" essentially justifies that the nonhuman realm is destined to be transformed into resources to serve humans. In this vein, a seminal paper, "The value of the world's ecosystem services and natural capital," became a widely cited study which people have used to place price tags on the nonhuman world.[22] Though the paper's underlying incentive is to call for us to care for the nonhuman realm, examining the nonhuman realm with a socio-economic lens implies and further justifies the attitude that "nature" is valuable only where it is transformed into capital.

Furthermore, because we frame the ecosystem as "services" for human society, the nonhuman realm becomes "replaceable" and "renewable." As an example, "no net loss" is a wetland preservation policy goal in the United States. Governments and agencies must balance "unavoidable" wetland loss—due to economic development—with wetlands restoration, mitigation, and reclamation efforts. The goal is to ensure that the total square-mile area of American wetlands either remains constant or increases. In a wetland assessment report for a traffic corridor improvement project, The Virginia Department of Transportation noted:

> The assessment of functional values is based on the understanding that certain wetlands are more valuable, offer more functions, and are of a higher quality than others. Wetlands with a higher standard of functional quality should be avoided as much as possible. Furthermore, they require a higher level of compensatory mitigation compared to impacts to poor quality wetlands with low functional values.[23]

We should praise the fact that the agency recognized the differences between wetland types. However, wetlands connect to broader landscape mosaics, so destroying a wetland in one place while restoring another at a different site would fundamentally alter two larger ecosystems. Increasing the total area does not mean an increase in the overall integrity of the larger ecosystem. The report quoted above demonstrates that agencies recognize the vagueness of the definition of "no net loss" and use extra caution when implementing the policy, by comparing different wetland types. Nevertheless, this line of narrative—placing values on wetlands based on how much functionality they offer—reveals a deeper issue in the "no net loss" policy; it ignores the internal complexity of a wetland as the habitat for countless nonhuman species, including animals, plants, fungi, and microbes. From the perspective of a wetland species, such destruction is irreversible, and the restoration of a separate piece of wetland in no way makes amends for the loss of a home.

Anthropocentrism manifests itself in landscape planning and design. Since Olmsted, the profession has always asked nonhumans to serve human society. Ian McHarg introduced systemic ways to design according to nature's instrumental values. Equipped with updated ecological concepts, the next generation of landscape architects argues for "landscape urbanism"—a new framework of urban growth that revolves around landscapes, rather than infrastructures and buildings. Landscape urbanism should be recognized for its role in expanding the boundaries of what landscape design may accomplish, despite critics who call it a mere rebranding strategy. It should also be praised for its effort to blur the discipline boundaries by turning landscape design into a model for other professions, such as urban planning, architecture, real estate, law, and civil engineering. However, landscape urbanism is essentially an update of the McHargian approach and bears similar critiques; it is "a dynamic outcome of ecological determinism *plus* economic determinism."[24] In landscape urbanism narratives, the ecological function is deeply integrated with economic growth.

New York City's High Line is an example of landscape urbanism. It expanded landscape architectural practices to urban infrastructure renovation by transforming an abandoned elevated railway in West Manhattan into a crowded urban park. Indeed, the High Line inspired numerous railway-line renovations in cities worldwide. Many of these cities saw potential, not in the urban habitats they provide for nonhuman species, but in the consequences of doing so: economic growth, increased property values, and thus tax revenues. According to The New York Times and the New York City Economic Development Corporation, between 2003 and 2011, the property values of residences near the High Line grew 103%. Landscape urbanism thus exploited and instrumentalized an ugly ramification of public parks, rebranding it with an economic justification. Moreover, even though "landscape urbanism" reflects the humans-nature dichotomy by exploring urban ecology and hybrid urban-natural systems, ironically, it reinforces this dichotomy by asking what nonhumans can further provide to serve human society.

Recognizing the intrinsic values of nonhumans is not a new idea. "Deep ecology" emerged as a philosophical concept in the 1970s. Norwegian philosopher Arne Naess coined the term to recognize the intrinsic values of other species, and the whole of the ecosystem, regardless of their instrumental values to human society.[25] This argument posits a non-hierarchical, flat ontology, underpinning many contemporary political ecology and environmental ethics arguments. Along the same lines, concepts such as *ecosophy* and *ecophronesis* ask questions of how to wisely think and act. These ideas posit that the survival of humanity requires the co-flourishing of the more-than-human whole.[26] They turn the concept "humans transform and use nature" into "co-flourishing of the more-than-human whole." This transition cultivates a sense of care and moral imperative that differ from many environmental efforts. We protect nonhuman species not because they are useful to us, either immediately or in the foreseeable future, but because human survival depends on the co-flourishing of both human and nonhuman realms.

Thus, the "nonhuman turn" becomes the theme of the third wave of the conceptual shift. Scholars take "nonhuman agency" as a conceptual framework with which to re-think nature. Such models as actor-network theory (ANT), assemblage thinking, object-oriented philosophy, and vital materialism have been discussed across disciplines, including landscape architecture. Despite the distinct metaphors deployed in these ideas, nearly all remove humans from the source of agency and hold that human agency is distributed across a network of human and nonhuman, living and nonliving things. In feminist materialist Jane Bennett's words,

> What this suggests for the concept of agency is that the efficacy of effectivity onto which that term has traditionally referred becomes distributed across an ontologically heterogeneous field, rather than being a capacity localized in a human body or in a collective produced (only) by human efforts.[27]

In contemporary design and planning disciplines, a "nonhuman turn" mirrors intellectual development within humanities and social sciences. "Multispecies coproduction" emerged as a concept in urban planning theory. It drew ideas from multiple intellectual traditions such as feminist materialism; "more-than-human" geographies; multispecies ethnography; political ecology; environmental humanities; and associated fields, and calls urban planners and designers to recognize that urban futures must be co-produced by and for both human and nonhuman agents. It challenges planners and designers to expand their moral, ethical, and political considerations to include the nonhuman realm, and in such a way as to cultivate relations that may nurture multispecies co-flourishing.[28]

To a certain extent, landscape architects have already practiced a coproduction mentality and posthumanist ethics. For example, when describing the oyster reef project in New York City, landscape architect Kate Orff rendered a shared urban public space literally co-produced by oysters, with a human culture built around them. The designers envisioned a landscape mosaic, which they called "oyster-tecture," to provide a habitat for oysters to stabilize and clean the water of New York Harbor. Over time, a new shared public space may be created by hardworking oysters and socio-cultural practices such as aquaculture and oyster harvesting. To illustrate the coproduction of the urban environment, Orff began with the statement that "New York was built on the backs of oystermen [sic], and our streets were literally built over oyster shells."[29] To describe her work, Orff notes,

> we have forgotten our relationship with the plants and animals that live alongside us and the dirt beneath our feet. And so, how I see my work contributing is [...] trying to literally re-imagine these connections and physically rebuild them.[30]

Orff also used anthropomorphic tones to discuss oysters as working creatures that clean water. Here, anthropomorphism is not necessarily anthropocentrism, but a sense of posthumanism. As Bennett argues, anthropomorphism "can catalyze a sensibility that finds a world filled not with ontologically distinct categories of beings (subjects and objects) but with variously composed materialities that form confederations."[31]

The third wave of reflection has given rise to a highly reflexive model of nature undergirded by an epistemological reconfiguration with a new mode of knowledge production. The autonomy of the nonhuman realm is irreducible to human discourse, and thus incomprehensible to any human-constructed model. However, to survive, we as humans constantly construct models to represent our observed world, and, in turn, the models decide how we choose to interact with the nonhuman realm. The coproduction thesis holds that the "ways in which we know and represent the world (both nature and society) are inseparable from the ways in which we choose to live."[32] Human knowledge about the nonhuman realm is always situated; it results from coproduction by both humans and nonhumans. The reason why we are who we are today is inseparable from the nonhuman realm in which we choose to engage. From this point of view, there is no pristine nature at all, because, from the beginning, nature is a result of coproduction, both conceptually and physically.

As a result of coproduction, the environment contains two senses. The first comprises the examples laid out by urban ecologists, landscape architects, and environmental historians. Humans are not the only actors to construct the environment. Other species, too, physically alter the shared environment. The second sense is knowledge production. In a traditional, humanist, epistemological framework, human exceptionalism leads to a belief that humans are the only ones who can construct models of self, others, and the system in which we are embedded. In contrast, the coproduction thesis begins with the premise that nonhumans are also able to construct models about themselves, humans, and their version of the systems in which they are involved. Humans and nonhumans co-evolve and co-produce each other, and the environment is a result of human and nonhuman coproduction.

The third wave of reflection also carries significant ethical implications. How we represent the nonhuman realm is inseparable from the model we elect to engage with it; therefore, we must cultivate those that help us recognize nonhuman agency, the co-evolution of humans and other species, and the coproduction of the environment.

Notes

1 Kingsland, "Conveying the Intellectual Challenge of Ecology."
2 Hill, "Shifting Sites."
3 Barbour, "Ecological Fragmentation in the Fifties."
4 Simard et al., "Mycorrhizal Networks."
5 Egerton, "History of Ecological Sciences, Part 54."
6 See a chart at https://commons.wikimedia.org/w/index.php?curid=479979. Last accessed: April 29, 2024. This chart was created using data from scientific papers.
7 Hill, "Shifting Sites," 131.
8 Gunderson and Holling, *Panarchy*.
9 Ellis, "Ecology in an Anthropogenic Biosphere."
10 McHarg, *Design with Nature*.
11 McHarg, 1.
12 Reed and Lister, *Projective Ecologies*.
13 The Structure of Scientific Revolutions: 50th Anniversary Edition.
14 Barbour, "Ecological Fragmentation in the Fifties," 249.
15 Barbour, "Ecological Fragmentation in the Fifties."
16 Cronon, *Uncommon Ground*.
17 Cronon, "The Trouble with Wilderness: Or, Getting Back to the Wrong Nature"; Slater, "Amazonia as Edenic Narrative."
18 Spirn, "Constructing Nature: The Legacy of Frederick Law Olmsted," 111.
19 Meyer, "Sustaining Beauty. The Performance of Appearance"; Meyer, "Values in Landscape Architecture."
20 Cronon, "The Trouble with Wilderness: Or, Getting Back to the Wrong Nature," 89.
21 Spirn, "Constructing Nature: The Legacy of Frederick Law Olmsted," 113.
22 Costanza et al., "The Value of the World's Ecosystem Services and Natural Capital."
23 VDOT, "I-81 Corridor Improvement Study: Wetlands and Water Resources Technical Report."
24 Steiner, "The Ghost of Ian McHarg," 149.
25 Naess, "The Shallow and the Deep, Long-range Ecology Movement. A Summary."
26 Naess, "From Ecology to Ecosophy, from Science to Wisdom"; Xiang, "Ecophronesis: The Ecological Practical Wisdom for and from Ecological Practice."
27 Bennett, *Vibrant Matter: A Political Ecology of Things*, 23.

28 Houston et al., "Make Kin, Not Cities! Multispecies Entanglements and 'Becoming-World' in Planning Theory."
29 Orff, "Transcript of 'Reviving New York's Rivers – with Oysters!'" https://www.ted.com/talks/kate_orff_reviving_new_york_s_rivers_with_oysters/transcript. Last accessed: April 29, 2024.
30 Orff.
31 Bennett, *Vibrant Matter: A Political Ecology of Things*, 99.
32 Jasanoff, "The Idiom of Co-Production."

Bibliography

Barbour, Michael G. "Ecological Fragmentation in the Fifties." In *Uncommon Ground: Rethinking the Human Place in Nature*, edited by William Cronon, 233–55. New York: W. W. Norton & Company, 1996.

Bennett, Jane. *Vibrant Matter: A Political Ecology of Things*. Durham, NC: Duke University Press, 2010.

Costanza, Robert, Ralph d'Arge, Rudolf De Groot, S. Faber, Monica Grasso, Bruce Hannon, Karin Limburg, et al. "The Value of the World's Ecosystem Services and Natural Capital." *Nature* 387 (1997): 253–60.

Cronon, William. "The Trouble with Wilderness: Or, Getting Back to the Wrong Nature." In *Uncommon Ground: Rethinking the Human Place in Nature*, edited by William Cronon, 1st ed., 69–90. New York, London: W. W. Norton & Company, 1996.

———, ed. *Uncommon Ground: Rethinking the Human Place in Nature*, 1st ed. New York: W. W. Norton & Company, 1996.

Egerton, Frank N. "History of Ecological Sciences, Part 54: Succession, Community, and Continuum." *The Bulletin of the Ecological Society of America* 96, no. 3 (2015): 426–74. https://doi.org/10.1890/0012-9623-96.3.426.

Ellis, Erle C. "Ecology in an Anthropogenic Biosphere." *Ecological Monographs* 85, no. 3 (2015): 287–331. https://doi.org/10.1890/14-2274.1.

Gunderson, Lance H., and C. S. Holling, eds. *Panarchy: Understanding Transformations in Human and Natural Systems*, 1st ed. Washington, DC: Island Press, 2001.

Hill, Kristina. "Shifting Sites." In *Site Matters: Design Concepts, Histories, and Strategies*, edited by Carol Burns and Andrea Kahn, 131–55. New York: Routledge, 2015.

Houston, Donna, Jean Hillier, Diana MacCallum, Wendy Steele, and Jason Byrne. "Make Kin, Not Cities! Multispecies Entanglements and 'Becoming-World' in Planning Theory." *Planning Theory* 17, no. 2 (May 2018): 190–212. https://doi.org/10.1177/1473095216688042.

Jasanoff, Sheila. "The Idiom of Co-Production." In *States of Knowledge : The Co-Production of Science and the Social Order*, edited by Jasanoff Sheila, 1–12, 2004.

Kingsland, Sharon. "Conveying the Intellectual Challenge of Ecology: An Historical Perspective." *Frontiers in Ecology and the Environment* 2, no. 7 (2004): 367–74. https://doi.org/10.1890/1540–9295(2004)002[0367:CTICOE]2.0.CO;2.

Kuhn, Thomas S. *The Structure of Scientific Revolutions: 50th Anniversary Edition*. Chicago, IL: University of Chicago Press, 2012.

McHarg, Ian L. *Design with Nature*, 25th ed. New York; Chichester; Brisbane; Toronto; Singapore: Wiley, 1969.

Meyer, Elizabeth K. "Beyond 'Sustaining Beauty.'" Edited by Ann Komara. *Values in Landscape Architecture: Finding Center in Theory and Practice* 71, no. 1 (January 2, 2017): 125–27. https://doi.org/10.1080/10464883.2017.1260964.

———. "Sustaining Beauty. The Performance of Appearance: A Manifesto in Three Parts." *Journal of Landscape Architecture* 3, no. 1 (March 2008): 6–23. https://doi.org/10.1080/18626033.2008.9723392.

Naess, Arne. "From Ecology to Ecosophy, from Science to Wisdom." *World Futures* 27, no. 2–4 (November 1, 1989): 185–90. https://doi.org/10.1080/02604027.1989.9972135.

———. "The Shallow and the Deep, Long-range Ecology Movement. A Summary." *Inquiry* 16, no. 1–4 (January 1, 1973): 95–100. https://doi.org/10.1080/00201747308601682.

Orff, Kate. "Transcript of 'Reviving New York's Rivers -- with Oysters!'" 2010. https://www.ted.com/talks/kate_orff_reviving_new_york_s_rivers_with_oysters/transcript.

Reed, Chris, and N.M.E. Lister. *Projective Ecologies*. Actar D, 2014.

Simard, Suzanne W., Kevin J. Beiler, Marcus A. Bingham, Julie R. Deslippe, Leanne J. Philip, and François P. Teste. "Mycorrhizal Networks: Mechanisms, Ecology and Modelling." *Fungal Biology Reviews*, Hyphal networks: mechanisms, modelling and ecology, 26, no. 1 (April 1, 2012): 39–60. https://doi.org/10.1016/j.fbr.2012.01.001.

Slater, Candace. "Amazonia as Edenic Narrative." In *Uncommon Ground: Rethinking the Human Place in Nature*, edited by William Cronon, 114–31. New York: W. W. Norton & Company, 1996.

Spirn, Anne Whiston. "Constructing Nature: The Legacy of Frederick Law Olmsted." In *Uncommon Ground: Rethinking the Human Place in Nature*, edited by William Cronon, 1st ed., 91–113. New York; London: W. W. Norton & Company, 1996.

Steiner, Frederick R. "The Ghost of Ian McHarg." *Log*, no. 13/14 (2008): 147–51.

VDOT. "I-81 Corridor Improvement Study: Wetlands and Water Resources Technical Report," 2007. http://www.virginiadot.org/projects/resources/Wetlands_and_Water_Resources_final-pages1–88.pdf.

Xiang, Wei-Ning. "Ecophronesis: The Ecological Practical Wisdom for and from Ecological Practice." *Landscape and Urban Planning* 155 (2016): 53–60. https://doi.org/10.1016/j.landurbplan.2016.07.005.

5

POSTHUMANISM, CO-PRODUCTION, AND ASSEMBLAGE

In the American context, the conceptualization of nature and technology as two kingdoms of force was deeply intertwined with the history of colonization, racism, and capitalism, creating interlocking stories of how humans use technology to transform nature into habitable landscapes at the expense of numerous humans and nonhumans. Such reasoning undergirds contemporary mainstream environmental narratives and fuels numerous uncritical environmental practices, which aim to protect the nonhuman realm but ironically further exploit those outside their mental models. Human exceptionalism is the fundamental premise based on which the two kingdoms of force could be articulated. A boundary between nature and technology was a convenient cover story for human agency and control.

To erase the illusory boundary and bypass the transformation formulas requires remapping the relationships between humans, nature, and technology within a posthumanist framework. As I have illustrated, to study technology and nature, in the end, is to study what it means to be human. In the discourse of nature, scholars have realized that the concept of "nature" fails to capture the autonomy and wildness of the nonhuman realm, and deploying this concept unwittingly reduces the nonhuman realm to a set of human discourses. In response, twenty-first-century intellects have started to use *nonhuman agency* and *multispecies co-production* to study the nonhuman realm. Regarding technology, historians and philosophers have realized that technology as an all-encompassing concept has prevented them from recognizing co-production between social and technological systems, and between human and technological artifacts. The two streams of study converged onto a posthumanist framework, which asserts that what we thought to be human agency has always been

DOI: 10.4324/9781003320852-7

distributed across a network of entities, including other species, tools, and systems.

What, then, is the implication of posthumanism in how we conceptualize nature and technology? What do these two terms denote rather than two kingdoms of force? And how do we visualize this "network of things?"

To answer these questions, I will rely on *assemblage* as a metaphor. Philosophy aims to develop metaphors to help analyze complex phenomena, such as nature and technology. This is less about truth and more about effectiveness, as well as the visual image the term renders. Aside from assemblage, metaphors such as actor-network and nested systems also convey the image of a "network of things." I choose assemblage as a metaphor not because assemblage theory describes the ultimate truth of reality, but because it is a useful metaphor for framing what technology and nature entail in a posthumanist framework.

First explored by philosopher Gilles Deleuze and psychoanalyst Félix Guattari in *A Thousand Plateaus* (1980), assemblage supplies a conceptual framework for analyzing social complexity, emphasizing fluidity, exchangeability, and multiple functionalities.[1] Philosopher Manuel DeLanda has updated and developed these insights into a more robust assemblage theory.[2] Assemblage thinking can be observed in the work of many posthumanist thinkers, such as feminist materialist Jane Bennett, STS scholar Bruno Latour, and object-oriented philosopher Levi Bryant. Broadly speaking, an assemblage is a multiplicity comprising many heterogeneous components, which themselves are also assemblages. The relationships between these components are not fixed; they can be "unplugged" and replaced with other parts or "plugged" into other assemblages. An assemblage has "emergent properties" that are irreducible to any of its parts.[3] Assemblage thinking is an ontological framework that gives us the flexibility to analyze across materially different entities, from mineral deposits and agricultural land to governmental organizations, policies, and laws. An assemblage can contain components from conceptually quite different categories, such as nature and technology.

Over the years, landscape architects have attempted to incorporate assemblage thinking in understanding landscapes as the result of the interactions of more than human agents.[4] We can imagine a diagram of a nested assemblage regarding a landscape (Figure 5.1).

When we observe this assemblage, we must be able to freely zoom in and out across scales. Zooming in, we observe humans, and nonhuman species, including plants, animals, and machines working on the land. We may zoom in closer to these entities, which are also assemblages composed of other components; a human eyebrow contains an ecosystem of microbes. Eventually, we should be able to zoom to the molecular level, visualizing water molecules as assemblages of hydrogen and oxygen atoms, which are themselves assemblages of electrons and quarks. Of course, with the development

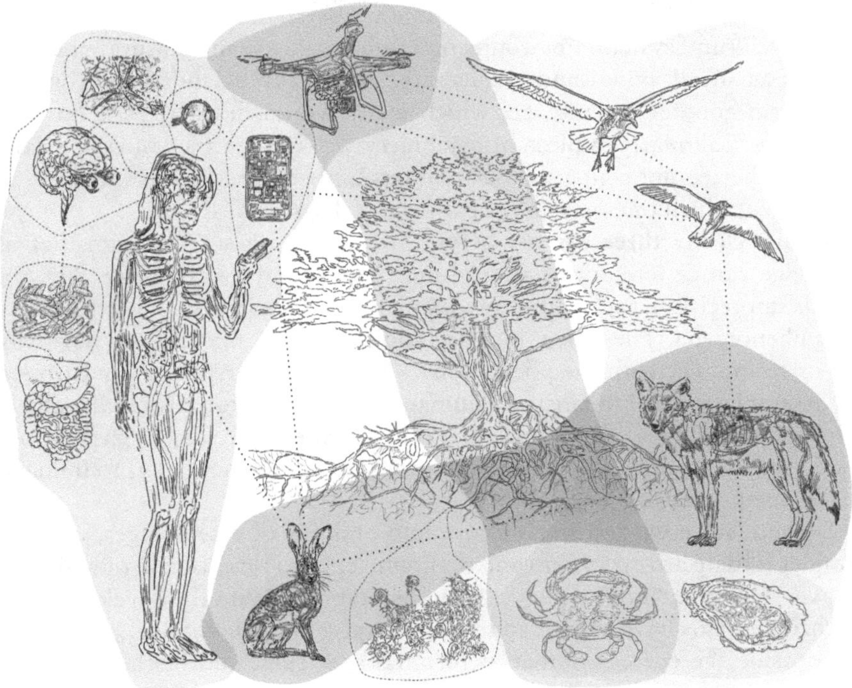

FIGURE 5.1 Networks of assemblages in a co-produced environment. Illustration by George McCracken.

of physics, other theoretical frameworks, such as string theory and quantum physics, may explain how we want to analyze these assemblages. The point here is that, on a micro-scale, categories such as human or nonhuman, nature or technology, living or nonliving are no longer relevant, because we are all composed of vibrant materials. These materials interact with and co-produce each other to form assemblages, which then take on emergent properties irreducible to any of their parts. These emergent properties are the results of co-production—for example, our ability to model nature and technology is a result of a long-term co-evolution with other nonhumans. The models make sense only within these interactions or within the assemblages which humans form with nonhumans. In other words, an assemblage must exist within another assemblage in order to act; an assemblage acts because it is made up of heterogeneous parts. As Bennett puts it, "[B]odies enhance their power in or as a heterogeneous assemblage."[5]

When we zoom out, a landscape assemblage is connected to a broader network of ecosystems and duly becomes a working component of other assemblages. Even though assemblage thinking is associated with new materialism, an assemblage need not be physical. For analytical purposes, a landscape may be temporarily "unplugged" from an ecosystem and "re-plugged" into a social-cultural system. Environmental policies, laws, and cultural practices, as well as social values and norms, are attached to a landscape. Together, they form another assemblage, which many would describe as a "cultural landscape." Similarly, a piece of machinery working on the landscape exists within other components, such as markets and engineering labs, as new assemblages, which one may call socio-technical systems.

I must clarify three caveats before proceeding. First, the claim that assemblage can denote any analytical unit may sound overly vague. Yet, due to this imprecision, the claim provides maximum flexibility to interpret complex phenomena. These new interpretations, in turn, may inspire new ways for us to interact with the phenomena. Once we understand a landscape as an assemblage—of humans, nonhumans, environmental laws and policies, intelligent machines, algorithms, and other entities—we may bypass the aestheticized image of the landscape and begin to imagine new ways to influence landscape processes.

Second, since we can freely zoom into and out of assemblages, one may wonder if there are maximum and minimum ends to this domain of scale. The answer would be "no," with a caveat of suitability. With the development of science, humans have expanded their perceptive spectrum. Some phenomena are outside the current human perceptual spectrum, and neither register a signal nor make meaningful impacts as information, but this need not mean they will remain meaningless in the future. For instance, until fairly recently, we did not know of mycorrhizal networks and their functions within forests, but these networks have become a frontier of ecological research and landscape design. The caveat would be that although we move freely across scales, that transition does not imply that we must analyze a phenomenon across all scales, because certain scales of analysis will not pertain to the question asked. For example, at least for now, it is less meaningful to analyze a landscape on a molecular scale.

Third, and most importantly, assemblage theory is less a theory than a mode of analysis. It claims no truth about the world. Instead, it is merely a lens and framework we intentionally adopt to interact with the world. The co-production thesis posits that the ways in which we understand the world cannot be separated from how we choose to live in it. Consequently, assemblage thinking is an intentional choice if one wants to view a world without boundaries between categories such as nature and technology. It is nearly a type of belief, as exemplified in Jane Bennett's "litany-like Nicene Creed" at the end of her *Vibrant Matter*:

> I believe in one matter-energy, the maker of things seen and unseen. I believe that this pluriverse is traversed by heterogeneities that are continually *doing things*. I believe it is wrong to deny vitality to nonhuman

> bodies, forces, and forms, and that a careful course of anthropomorphization can help reveal that vitality, even though it resists full translation and exceeds my comprehensive grasp. I believe that encounters with lively matter can chasten my fantasies of human mastery, highlight the common materiality of all that is, expose a wider distribution of agency, and reshape the self and its interests.[6]

Using the assemblage diagram, *nature*, *technology*, *landscape*, and *human* become concepts which are co-produced by heterogeneous assemblages. These concepts can be represented as amoeba shapes circling these assemblages in a heterogeneous field. Again, "amoeba shape" offers yet another metaphor, because the boundaries are fluid and constantly shifting. As we have seen, the meaning of these terms depends highly on the observer's perspective and the time in which they lived.

As I have illustrated, constructivists viewed nature as a socially constructed idea laden with culturally specific values and norms, whereas ecologists perceived nature as a research object to be examined under a scientific apparatus. In the early twentieth century, nature meant a static nonhuman realm with human disturbance. Toward the twenty-first century, however, nature took on the meaning of "dynamic and continually changing systems," with humans as major factors. Similar examples may be found in the history of technology. The concept had pointed to a branch of study of the mechanic arts in the late nineteenth century, but by the early twentieth century began to denote unprecedented, progressive human power. The term's meaning expands and shifts as society assimilates new technological artifacts and innovations. The term *landscape* does the same. Its meaning transitioned from the early *Landschaft*—people working on the land—to an aesthetically pleasing view. Within the profession of landscape architecture, landscape may mean an assemblage of human and nonhuman actors that co-produce the shared environment.

Most importantly, the narratives of nature and technology in America reveal that the transformation formula required a wielder of the powers of nature and technology. The early settlers needed a concept of *human* that was constructed as the source of agency, to tell the second creation story, which was, from our current perspective, a muddled interaction and co-production of human and nonhuman assemblages. As posthumanist proponent Katherine Hayles notes, "Mastery through the exercise of autonomous will is merely the story consciousness telling itself to explain results that actually come about through chaotic dynamics and emergent structures."[7] In North America, the human in the transformation formula was simply a character which early colonists invented to attribute agency and to make sense of perceived causality—unlimited technological power plus natural abundance yielded the middle landscape. Thus, the concept of human in the transformation formula is merely a version of the mapping of other concepts—a specific arrangement of knowledge. As Michel Foucault famously puts it at the end of *The Order of Things* (1966), the appearance of humans was nothing but "the effect of a change in the fundamental arrangements of knowledge...If those arrangements were to disappear as they appeared...then one can certainly wager that man would be erased, like a face drawn in sand at the edge of the sea."[8]

Assemblage thinking is a posthumanist idea, since *human* in this diagram is framed as distributed systems. Human capacities, which we thought originated from the human body, are achieved through a network of nonhuman assemblages. Even the human body itself is a heterogeneous assemblage of many nonhuman and nonliving entities. As a consequence, the boundary of *human* becomes as ambiguous and undetermined as other terms—nature, technology, and landscape.

Unlike approaches such as coupling, hybrid, and cyborg, that attempt to dissolve the boundaries between these categories by joining the opposite sides, assemblage thinking does not eliminate the illusory boundary at all. Instead, it renders an ontological framework in which drawing boundaries becomes less important, if not completely useless. In this assemblage diagram, the categories become assemblages, with overlapping, constantly changing amoeba shapes, and any static images of landscape, humans, technology, and nature become ephemeral and contingent to the context of analysis.

With the assemblage diagram, there is no such entity as "the environment" but always "an environment of." The term *environment* implies the exterior of these assemblages; the term is akin to its original sense, as in "environs," denoting the exterior of a system. If one draws a boundary around certain assemblages and names them human, then what exists outside this boundary is the environment of the temporary "human assemblage" to which we attribute agency. Moreover, because the boundary of the "human assemblage" constantly shifts, the environment of the "human" becomes ambiguous. When someone uses a smartphone's online map to navigate through a city, that phone, the online map database, the navigation algorithm, the human body, and many unknown actors form a temporary working "human assemblage" to perform the task of way-finding. Whatever temporarily exists outside this system of analysis becomes a temporary environment of this "human assemblage."

From this vantage, I argue that a diagram of assemblage should replace the transformation formula to explain how the environment works. Concepts of nature, technology, landscape, and human point only to temporary working assemblages with ambiguous boundaries. The tensions between nature and technology, as well as the garden-machine motif, are dissolved within the field of heterogeneous assemblages.

Notes

1 Deleuze and Guattari, *A Thousand Plateaus.*
2 DeLanda, *Assemblage Theory.*
3 DeLanda.
4 Prominski, "Andscapes"; Davis, "Landscapes and Instruments."
5 Bennett, *Vibrant Matter: A Political Ecology of Things*, 23.
6 Bennett, 122.
7 Hayles, *How We Became Posthuman: Virtual Bodies in Cybernetics, Literature, and Informatics*, 288.
8 Foucault, *The Order of Things: An Archaeology of the Human Sciences*, 422.

Bibliography

Bennett, Jane. *Vibrant Matter: A Political Ecology of Things*. Durham, NC: Duke University Press, 2010.

Davis, Brian. "Landscapes and Instruments." *Landscape Journal* 32, no. 2 (2013): 293–308.

DeLanda, Manuel. *Assemblage Theory*. Speculative Realism. Edinburgh: Edinburgh University Press, 2016.

Deleuze, Gilles, and Félix Guattari. *A Thousand Plateaus: Capitalism and Schizophrenia*. Minneapolis: University of Minnesota Press, 1987.

Foucault, Michel. *The Order of Things: An Archaeology of the Human Sciences*. 1st American ed. World of Man. New York: Pantheon Books, 1970.

Hayles, Katherine. *How We Became Posthuman: Virtual Bodies in Cybernetics, Literature, and Informatics*. Chicago, IL: University of Chicago Press, 1999.

Prominski, Martin. "Andscapes: Concepts of Nature and Culture for Landscape Architecture in the 'Anthropocene.'" *Journal of Landscape Architecture* 9, no. 1 (January 2, 2014): 6–19. https://doi.org/10.1080/18626033.2014.898819.

PART II

Posthumanism, the Environment, and Intelligent Machines

6

SEARCHING FOR NONHUMAN AGENCY

Posthumanism rose as a movement and entered other fields, including landscape architecture, in the twenty-first century. Many landscape scholars found that the posthumanist proponents' urge to cultivate empathy toward the nonhuman realm aligned well with the field's ethical concerns. However, the posthumanists' wishful thinking is incompatible with *design*. Posthumanism frames human agency as acentric and distributive across a network of assemblages, but *design* in its traditional senses has been envisioned based on a clear human agent and centralized agency. How might we re-conceptualize design within a new ontological framework? What does a design strategy look like if human agency is understood as distributive? How do we include nonhuman agency in a co-produced future? If we introduce posthumanism into landscape architecture, these are the central questions.

To address them, I will investigate how different posthumanist ideas conceptualize nonhuman agency within assemblages so that we can reconfigure the concept of design within a posthumanist framework. This chapter presents a group of representative posthumanist ideas, including actor-network theory (ANT), vital materialism, object-oriented ontology (OOO), and ontology of machines and media. These ideas are presented as a progression from human agency to agency of material and assemblage, and from post-attribution of agency to searching for an inaccessible "surplus" in objects.

To discuss nonhuman agency, one must resolve an inherent tension in this proposition, the tension between "the stubborn reality of individuation and the essentially distributive quality of their affectivity."[1] In the assemblage framework, there must never be a source or a center from which agency is generated; yet, we must daily confront individual objects and systems, and when discussing agency, we still need to rely on individualistic terms such as

DOI: 10.4324/9781003320852-9

"nonhuman" or "human" agency. Different posthumanist concepts, such as actor-network, assemblage, object, machines, and vibrant matters, attempt to provide new vocabularies and metaphors to resolve the tension between individuation and distributed effectiveness. In ANT, the actor-network acts as a metaphor for how nonhuman actors participate in social assemblages and diversify human agency. Similarly, materialist feminism scholar Jane Bennett turns to "mater-energy" as a concept that cuts across living and nonliving things, and thus, through a type of monism, dissolves individuals within a field of vibrant matters.[2] On the other hand, object-oriented philosopher Levi Bryant reinforces individuals and uses the concept of "gravity" and "gravitational field" to explain how objects influence and modify each other's agency through mediation.[3]

These posthumanist ideas, except perhaps for several versions of ANT, all share a concept of "surplus," which denotes unknowable aspects of assemblages beyond any form of human access. No matter how many nonhumans we summon to attribute agency, this "nonhuman agency" is forever a perceived effectiveness in human terms; at their best, nonhumans can only diversify and hybridize human agency. This is why OOO proponent Graham Harman completely bypasses the concept of agency to search for the inaccessible "surplus" in objects. Thus, OOO argues that objects are withdrawn from direct access, and aesthetic experiences and metaphors serve as indirect access to real objects.[4] We may use the OOO framework to re-articulate design practice as a way to constantly create metaphors—unusual connections between objects—so that different objects or assemblages can exploit each other's surplus and produce shared effects.

Actor-Network Theory

In the English context, ANT often suggests a method of analysis developed by Michel Callon, Bruno Latour, and John Law over the past several decades in the field of science technology and society (STS). Chapter 3 above introduced ANT in the context of the social construction of technology (SCOT) movement in the 1980s, and STS scholar Weibe Bijker regarded ANT as an approach with which to analyze how technology was made.[5] Bijker's categorization was a strict reading of ANT in the SCOT context. Rather than a version of strict constructivism, the true implication of ANT lies in its ability to counter radical social constructivism by including nonhumans as actants that shape society. We may understand ANT as a corrective framework that emerged in the late 1980s to reflect on the prevailing radical constructivism of the time. The concept of ANT suggests that anything in the social and natural world exists only in a continually shifting network and that there is no exterior social force except what exists in the network acting and being acted upon; "entities are constituted by the relations that they are enrolled in."[6]

Outside the STS context, ANT was introduced to the broader intellectual debate by Bruno Latour, and, in the twenty-first century, it has become a widely used framework to analyze nonhuman agency. For instance, Jones and Cloke have chosen trees as examples to consider how nonhuman agency manifests in the actor-network.[7] They proposed four ways in which nonhuman agency can be considered. The first is "agency as routine actions." Trees have agency due to a series of routine processes, such as reproducing, fruit-bearing, and colonizing. Second, nonhuman agency can be considered a transformative action. Trees autonomously seed themselves and grow in unexpected places and forms, and "when remixed with the social aspect, these actions can have creative transformative effects."[8] Third, nonhuman agency can be considered as purposive action. To account for the intentionality that is often associated with human agency, but without falling into a form of reductionist essentialism, Jones and Cloke have coined the term "purposive agency" to describe the process whereby a tree executes a plan inscribed in its DNA. The fourth is agency as non-reflexive action. The notion of human agency is commonly built around creative or reflexive actions that require agents to set goals, reflect on intentions, prioritize, and rework goals. Trees exercise a type of non-reflexive agency, through hybridizing human creativity. They "engender affective and emotional responses from the humans who dwell amongst them—to contribute to the haunting of place via exchanges between the visible present and the starkly absent in the multiple and incomplete becoming of agency."[9]

The above analysis displays a struggle: in attempting to attribute agency to trees, the authors found themselves in an awkward position where they were unable to locate intentionality, reflexivity, or creativity—the concepts associated with human agency—in trees; thus, they were forced to subjugate trees to human interpretation in an actor-network. What nonhuman agency genuinely means in their analysis becomes the roles which, through narratives, the narrators assign to nonhuman actants. This struggle reflects a paradox in employing ANT as a framework for conceptualizing nonhuman agency; this paradox is a manifestation of the tension between individuation and distributed effectiveness. In order to recognize the contribution of the nonhuman realm to the shaping of the social, ANT posits a flat ontology, an equal treatment for both human and nonhuman, or, as many ANT scholars would call it, "generalized symmetry." This version of flat ontology, which treats human and nonhuman on equal terms, allows ANT to recognize "the agency of nonhumans as an essential element in how the natural and the social flow into one another."[10] However, ANT "has rejected the nonhuman/human distinction" and renders a "hybrid collectif" in which agency is relational and distributive.[11] Thus, ANT "[subjugates] the specific importance of individual actors within networks, in order to focus on the multiplicity of mutually constitutive and positioning 'actants' which together serve to hybridize agency."[12] In other words, neither "human agency" nor "nonhuman agency" exists. Human and nonhuman "actants" form hybrids, which "are then seen as mobilized and assembled into associative networks in which agency represents the collective capacity for action by humans and nonhumans."[13] Essentially, ANT is a version of assemblage thinking, and it "sees agency as a distributed achievement, emerging from associations between human and nonhuman entities (the actor-network)."[14]

Still, ANT is a human-centered undertaking because of its roots in social constructivism. By the 1980s, social constructivism had advanced in the field of STS as a major framework of analysis to challenge the notion of "ready-made technology" in technological determinism, but the obverse is that it unavoidably presumed a "ready-made society." The innovative aspect of ANT was in regarding biological and technological nonhumans as actants that were an essential part of society. However, an actant or actor plays a role in a narrative only through its society-generated meaning. Therefore, to exercise agency, nonhumans must be subjugated to an actor-network created by human interpretation. In ANT's framework, nonhumans can never truly exercise agency outside the actor-network because they must be transformed into actants that function in a set of human narratives.

Posthumanist Cary Wolfe has analyzed a range of posthumanist thinkers. Perhaps because of ANT's humanist undertaking, Latour barely achieved a place in Wolfe's "posthumanist posthumanism" category, and nearly fell into "posthumanist humanism."[15] Thus, ANT is restricted by its original goal to "reassemble the social," and it presents society with both human and nonhuman actants. At its best, ANT redefines "social" by including nonhumanity as part of what we had thought to be humanity, but it never intended to ask the question, "What does nonhumanity mean?" To consider the nonhuman realm is, for ANT, an afterthought, not an element for which ANT was initially designed. We may update and develop the concept, as others have done, but over-interpretation only makes the already complex concept slipshod. Today, one must deal with "post-1999 ANT," "Latourian ANT," "more-than-Latourian ANT," and numerous interpretations from scholars and commentators.[16] One must also distinguish between "younger Latour" and "older Latour." With the passing of Bruno Latour in 2022, it would not be surprising if one soon must deal with postmortem Latourian ANT. Excessive interpretations of ANT have backfired since ANT was never intended as a theory, but was an analytical framework to be practiced rather than explained; excess articulations made it rigid and even useless. Instead, we must focus on the similarities and differences between its underpinning ontological framework and other posthumanist ideas and extract useful concepts to nurture the thesis of co-production of the environment.

Müller and Schurr observe similarities between ANT and assemblage thinking, in at least four aspects: both (1) "have a relational view of the world, in which action results from linking together initially disparate elements"; (2) "emphasize emergence, where the whole is more than the sum of its parts"; (3) "have a topological view of space, in which distance is a function of the intensity of a relation"; and (4) "underscore the importance of the socio-material...that the world is made up of associations of human and nonhuman elements."[17] Müller and Schurr also detect a difference between the two concepts. Even though assemblage thinking acknowledges the influence of an assemblage upon its components, through what DeLanda describes as "downward causality," any element in an assemblage may be "unplugged" from that assemblage and "re-plugged" into another.[18] In this view, nonhumans are not entirely constrained by the relations or the "actor-network" in which they are enrolled as actants, but they "always exhibit a surplus, something that is outside relations, and enables them to plug into other assemblages."[19]

Perhaps this "surplus" is the "nonhuman agency" for which Jones and Cloke searched in trees. However, ANT's underlying ontological framework cannot hold such a "surplus." To resolve the tension between individuation and distributed agency, ANT transforms nonhumans into actants whose roles are assigned only by the narrators; this "surplus" is overwhelmed by the roles they must play. Such ANT terms as "actants," "translation," and "free association" render it more of a linguistic analysis, which cannot account for physical interactions between the human body and other material bodies. For example, in Jones's and Cloke's analyses, the transformative agency of trees is more than a result of language and narration. When the root of a street tree cracks a hard pavement, so that pedestrians adjust their steps to bypass the obstacle, the roots' impact is physical and material. To a certain extent, contemporary landscape architects have explored transformative agency for decades. In SCAPE's oyster-tecture project, detailed in Chapter 4, the designer was well aware that the oysters participated in the project more diligently than their socio-cultural roles in wishful design narratives would have assumed; their ability to participate in hydrological and ecological cycles in the urban environment strengthens the urban ecology in constructing a new sort of public space. Is the "surplus" a type of material agency? This question might be better analyzed through Jane Bennett's vital materialism, which may be understood as a materialist's approach to ANT.

Vital Materialism

Bennett's thinking is instrumental for new materialism and political ecology. Bennett refuses to draw a clear boundary between ANT and assemblage thinking; she works instead with both ideas by jointly using their vocabularies. In her view, agency is confederate and distributive. Drawing from Spinoza's conative bodies, Bennett states that "bodies enhance their power in or as a heterogeneous assemblage," and "...the efficacy of effectivity to which that term [agency] has traditionally referred becomes distributed across an ontologically heterogeneous field, rather than being a capacity localized in a human body or in a collective produced (only) by human efforts."[20] Vital materialism acknowledges the network and confederate status of things; Bennett argues that "there was never a time when human agency was anything other than an interfolding network of humanity and nonhumanity."[21] She draws ideas from Francois Julien's reflection on a Chinese concept, *shi*, to formulate her discussion of the agency of assemblages. "*Shi* is the style, energy, propensity, trajectory, or élan inherent to a specific arrangement of things...*shi* names the dynamic force emanating from a spatio-temporal configuration rather than from any particular element within it."[22] A detailed exploration of *shi* is beyond the scope of this book, but we can, for now, understand *shi* as similar to "surplus," a type of emergent tendency or propensity due to a specific material arrangement.

In ANT's framework, human agency is the model for agency; the concept is associated with creativity, intentionality, consciousness, and free will—the concepts commonly used to describe human traits. There is still a liberal human subject who can exercise free will; from this vantage, ANT is not a posthumanist idea at all. Unlike ANT, vital materialism does not assume a clearly bounded human body, or any bodies. The human body "is material, and yet this vital materiality is not fully or exclusively human," and for vibrant materialism, the human is "an array of bodies, many different kinds of them in a nested set of microbiomes."[23] In vital materialism, nonhuman agency, or, to be more precise, material and assemblage agency, became the model of agency. This is a posthumanist idea, because terms such as *consciousness* became concepts invented by human assemblage to describe the observed effectiveness that emanates from the specific arrangement of materials we conceptualize as human. While ANT exploits nonhumans as actants to hybridize human agency, vital materialism redefines agency as an emergent property of material assemblage so that the distinctions between human and nonhuman, and even between living and nonliving, recede in significance. Humans, plants, animals, rocks, and soils are all created of matter that is "vibrant, vital, energetic, lively, quivering, vibratory, evanescent, and effluescent."[24] Bennett used these adjectives throughout her work, painting an active and lively world of vibrant matter, and a heterogeneous yet unified field of matter-energy.

Bennett has become widely read in the landscape architecture discipline over the past decade since her positive ontological framework and sensibility toward the biophysical world echo many of landscape designers' and scholars' beliefs. After reading *Vibrant Matter*, Brett Milligan, a landscape educator, averred that "we need a language for reading the landscape and a corresponding design sensibility with similar capacities for inclusion and complexity."[25] Indeed, over the past several years, the vocabularies developed in vital materialism have been incorporated into landscape design lexicons; many designers rely on these concepts to describe a designed landscape as the joint effort of diverse vibrant bodies.

However, it appears that *agency* as a term is unable to adequately describe that for which Bennett is truly searching. In Bennett's narrative, a ghost of "surplus" always exists. For example, toward the end of *Vibrant Matter*, Bennett begins to doubt whether to use *agencies* or *agents*. In one note, she writes, "[a]s I struggle to choose the right term, I confront a profound ambiguity in both terms regarding wherein lies the cause and wherein the effect."[26] Bennett wanted to use *agency* to denote an underlying cause—a mystery "surplus"—for an agent to act rather than the perceived effectiveness of the agent's action. In many places, Bennett altogether avoids the notion of agency. Instead, she relies on terms such as affectivity, efficacy, and effectiveness. Toward the end of her argument, Bennett turns to "matter-energy," a metaphysical term, to describe what the notion of *nonhuman agency* fails to capture.

At this point, it is clear that the notion of "surplus" has emerged as the concept for which ANT scholars and new materialists have been searching, and that the notion of "surplus" is different from the meaning which the word *agency* conveys. OOO questions this "surplus" directly, by arguing that objects are withdrawn from any form of access, thus highlighting that "surplus" exists outside our human conceptual framework.

Object-Oriented Ontology

Graham Harman began to use the term "object-oriented philosophy" (OOP) in approximately 1997 and publicly employed the term in 1999.[27] By 2009, he had been joined by three object-oriented philosophers: Ian Bogost, Levi Bryant, and Timothy Morton. Bryant coined "object-oriented ontology" (OOO) as a catch-all for their shared ideas. Though OOO has replaced OOP in many discussions, Harman has noted that he reverted to OOP because of increasing philosophical disagreement with Bryant.[28] Yet, in his 2018 book *Object-Oriented Ontology*, Harman returned with a full explanation of OOO and published his version of this somewhat controversial school of thought.[29]

Harman's OOO follows two premises. First, OOO shares a flat ontology with other posthumanist ideas, but it is the "flattest" version, because it claims that the human/nonhuman and thought/world distinctions in philosophy still attribute at least 50% of our attention to humans. In contrast, OOO seeks to produce a flat ontology in which all objects should be treated equally, and humans comprise only a fraction of a vast, mystical universe. Second, OOO begins by critiquing other philosophical ideas, utilizing a framework which Harman calls *anti-mining*. Thus, OOO claims that, to date, most philosophical analysis has reduced objects to two types of knowledge: *over-mining* and *under-mining*. Over-mining explains what an object does, reducing an object to its relationships with other objects. For OOO, ANT is a form of over-mining that reduces nonhuman objects to a set of human relations. On the other hand, under-mining explains what elements a thing is made of, thus reducing objects to their material components. For Harman, Bennett's vital materialism and assemblage thinking largely belongs to the under-mining category because this type of knowledge reduces objects to their components.

This critique of "dual-mining" sets up OOO's mission for philosophical inquiry: to ask about the third type of knowledge about objects, which lies beyond direct access and is always "withdrawn."[30] Harman relies on a diagram of the "Quadruple Object"—real object, real quality, sensual object, and sensual quality. This object-quality framework consists of two pairs, real and sensual, in both categories. Real objects and qualities cannot be accessed by any means, and interactions occur only through sensual objects and qualities, or the fictional images which objects present to each other. This inability to access lays the foundation for OOO's ontological framework, in which real objects and qualities are always hidden from access, thus "withdrawn."[31]

Furthermore, OOO has remained a controversial school of thought, partially due to Harman's straightforward rejection of other ways of thinking and his philosophical approach that emphasizes differences rather than similarities. Architecture scholar Neil Leach has harshly critiqued OOO, yet speaks highly of DeLanda and new materialism in a conference paper for ACADIA (Association for Computer-Aided Design in Architecture).[32] Leach's critique was both response and warning to the growing interest in and interpretations of OOO within the discipline of architecture and its adjacent fields, including landscape architecture. Leach focuses his critique primarily on two schools of thought that influenced Harman's OOO: the neo-Heideggerian approach and the ANT approach to technology. Harman developed OOO within his Ph.D. dissertation on Heidegger and his tool analysis. Leach, however, finds Heidegger's approach to technology problematic and argues that "[w]hat Heidegger fails to address...is the progressive way that we come to appropriate technology in general, and tools in particular, and absorb them within our horizon of consciousness"; what bothers Leach is that, for Heidegger, "technology is perceived as antithetical to what it is to be human."[33] Moreover, Leach raises an alarm regarding Latour and ANT because he finds it problematic to assume that objects act in social networks.[34] Leach's fundamental problem with OOO is its potential danger of ascribing "agency" to objects, especially technical objects such as digital tools, as if they could think and act.

Leach's critique represents a generally held misinterpretation of OOO, or at least, Harman's OOO. In Harman's framing, a human, as a real object, may interact with technical objects only through their sensual objects and qualities. In other words, what we thought to be human in Harman's framing has always held a proportion of technical objects within it. Harman's framework does not conflict with what Leach believes: that humanity can "absorb" technology. Moreover, finding problems in Heidegger is insufficient to deny OOO as a whole. In addition, like other scholars, Leach confronts multiple versions of ANT that exist in parallel, and it is simple to develop a misinterpretation of the complicated ontological framework of actor-network. As we have seen, in ANT, objects do not truly act, but instead join the actor-network as actants to hybridize agency. Leach misses that the nonhuman has always been involved in an interfolding network of assemblages, and agency has existed forever in hybridized forms. Over-interpreting a useful concept such as ANT will miss its similarity to assemblage thinking and new materialism. When Leach argues that humans "absorb" technologies, he essentially restates an ANT notion: ANT argues that technical objects become actants that "act" in human narratives.

Moreover, at its heart, Leach's argument is a humanist undertaking and does not venture enough to embrace posthumanism. Leach's concerns about attributing agency to tools reveal a potential terror evoked by a misreading of posthumanism as antihumanism—the displacement of humans by intelligent machines. Indeed, a more profound concern is that we may retreat to outdated technological determinism. However, these concerns are caused by bearing in mind a liberal human subject who can exercise free will; as far as one associates agency with this image of the human, one will feel troubled in embracing the agency of other assemblages. To alleviate these concerns, one must regard agency as distributive. We may realize that, for exercising agency, not only does it decenter the human from a privileged position, but it also decenters machines from any privileged position to initiate change. If there were anything to call "machine agency" (with which machines can initiate change), it would already be distributed in the socio-technical network.

Thus, OOO is often viewed as a counterpart to Bennett's vital materialism and assemblage thinking. If vital materialism is a version of monism, which dissolves boundaries between individuals and objects in a field of matter-energy, thus emphasizing relational agency, then OOO celebrates the autonomy of individual objects. In the essay "System and Things," Bennett for the first time expressed mixed feelings toward OOO's approach, which renders a muted nonhuman realm by deploying terms such as *withdrawn*.[35] In *Object-Oriented Ontology* (2018), Harman responded to Bennett with "the same mixture of sympathy and unease."[36] Clearly, they share deep empathy and attention to the life of nonhuman things, but Harman viewed their disagreement from three aspects:

> ...OOO does not view the world as a unified whole that is only secondarily broken up into individuals; it does not endorse the concept of matter at

all...OOO does not think the ultimate role of objects is doing, which for most of us can only count as a form of overmining.[37]

Although OOO might become an inspiring concept, Harman's reluctance to look for common ground with other posthumanist ideas made OOO a controversial school of thought. In fact, OOO shares similarities with other posthumanist ideas. First, Harman uses the concept of symbiosis to argue that objects do interact and form new objects that exist in their own right. This notion differs not at all from assemblage thinking, which posits that components can form a new assemblage that takes on emergent properties irreducible to any of its components, and thus the new assemblage exists in its own right. From this perspective, "withdrawn" might be understood as similar to "surplus" in assemblage thinking.

Second, Harman discusses how two real objects can interact with each other through the mediation of two sensual objects, or, to use OOO terms, "only by way of the fictional images they present to each other."[38] Harman's argument may be articulated by autopoiesis theory, an important concept in second-order cybernetics, developed by Chilean biologists Humberto Maturana and Francisco Varela. Chapter 9 will explore this concept in detail. For now, we need only know that an autopoietic system, such as a human, uses inputs to reproduce the organization of its components. Thus, when two autopoietic systems interact, they inevitably reduce each other to a set of "fictional images" through their system operations and self-organization processes.

Third, when Harman analyzes what knowledge means in OOO's framework, he argues that knowledge is about attempting to access real quality through sensual objects—the image exists in the human mind, but the real qualities of the sensual object cannot be accessed. To maintain the object-quality pair, the real qualities of the human beholder must step in to replace the real qualities of the sensual object. Thus, Harman argues, "the real qualities in knowledge come from the beholder rather than the sensual object of knowledge itself."[39] Harman then follows with a question: "In what sense does the beholder supply real qualities for a sensual object?"[40] In answer, he eventually turns to Kuhn's concept of paradigm to explain the ways in which real qualities of the beholder develop. However, if we eliminate the complexities within object-quality pairs, we find that Harman essentially argues for a version of radical constructivist epistemology stemming from the idea of autopoiesis: knowledge is constructed by the system to maintain its own system organization. The "real qualities of beholders" is just another way to say "system self-organization."

If there were a single difference that distinguishes OOO from other assemblage-oriented posthumanist thoughts, it would be the chicken-and-egg dilemma. While most assemblage ideas, including new materialism, assemblage theory, and ANT, begin with a pre-existing network or assemblage of things, and individuation is a *post hoc* attribution of agency by observers, OOO posits that "the world is home to preexistent unified entities that have individual shapes prior to being encountered by some observer."[41] The emphasis on individuals is why Harman and his fellow object-oriented philosophers seek to revive and deploy the stubborn notion of "object" as a metaphor to describe their ontological framework. However, this difference should not cause significant incompatibility between OOO and other posthumanist ideas, if we focus on the larger picture to look for the "surplus" and "real object" in objects and assemblages.

Ontology of Machine and Media

Levi Bryant presents an alternative version of OOO. Distinct from Harman's negative approach to modern philosophy and intellectual developments, Bryant focuses on similarities and develops a framework that shifts between separate schools of thought. As he wrote,

> [w]hat we need is a posthumanist framework that is able to synthesize the findings of the linguistic turn, Marxist thought, Foucaultian thought, media theorists...as well as the posthumanist thought of the ecologists, the new materialists, the actor-network theorists [...so that] we can begin to develop maps adequate to the political and ethical demands that face us today.[42]

Bryant does not shy from admitting his intention to deploy certain terms to replace others to develop a cohesive framework that accounts for different schools of thought. Bryant uses the term "machine" instead of "object" to embrace that which systems theory and second-order cybernetics offer. The term *machine* emphasizes the exchange of material and information flow between entities and the functional and operational relations between them. Bryant argues that machines are structurally open and operationally closed—terms directly borrowed from second-order cybernetics, autopoiesis, and systems theory. Moreover, Bryant uses "positive feedback" and "negative feedback" to explain how machines form structural couplings with each other. Any machine may function as media for another machine. He deploys the concept of "gravity" to replace "power" in other schools of thought, to denote "the way in which the structural openness, movement, and becomings of one machine are mediated by another machine."[43] To a certain extent, we can either understand that Bryant placed a posthumanist touch on second-order cybernetics and systems theory or say that he re-introduced systems thinking into OOO, which intended to move away from over-emphsized relational thinking found in the system metaphor in the first place.

In Bryant's framework, there is also a "surplus" in machines; as he argues, "...*machine itself* issuing certain imperatives on its designer that run away from the intentions of the designer."[44] Moreover, he names one category of objects as *dark objects*. They are

> thoroughly unrelated to other machines—most importantly, ourselves—that we would have no idea of their existence whatsoever [...] Like spirits or ghosts, it is therefore possible that any given assemblage is haunted by all sorts of dark objects that do not manifest themselves and that they have no effect on other machines of the assemblage whatsoever.[45]

Bryant acknowledges that this idea appears to be an outlandish metaphysical concept, but he reminds us "not to reduce the world to the machines that we happen to encounter in the world" and not to reduce a machine to its local manifestation.[46] To Bryant, a machine always has "a bit of darkness within it," and "[t]he domain of a machine's power is always greater than how it happens to locally manifest itself at any given point in time and under one particular set of circumstances of gravitational relations to other machines."[47] Bryant's "dark object" speaks to the "surplus" of assemblage thinking and the "real object" of Harman's OOO.

Apart from shared similarities, Bryant also provides a posthumanist account of agency. He argues that agency is a concept related to agent and posits two criteria for a machine to count as an agent. First, an agent is able to initiate action from within itself. Second, an agent must be able to choose whether or not to initiate an action.[48] From these two criteria, Bryant continues to argue that

> agency comes in a *variety of degrees*. Bacteria appear to have more agency than rocks insofar as they seem capable of initiating action from within themselves, whereas rocks cannot, while cats seem to have more agency than bacteria in that they are capable of choosing among a broader range of possible actions for themselves.[49]

At first glance, Bryant seemingly disassociates from the framing that agency is distributed across heterogeneous assemblages. Yet, Bryant's treatment supplies flexibility because it ties the notion of agency back to observation. Through his descriptions, we know he is aware that what he calls "agency" is, in fact, *perceived agency* from his perspective as an observer. This move provides more plasticity to account for different types of assemblages since as long as an observer or narrator circles an assemblage as an "individual," they can observe a certain degree of agency within it. Then, machines can aggregate and form new agential assemblages that take on different degrees of agency. Through this agential treatment, the notion of agency is no longer intertwined with "surplus" and "dark object" as an essential quality of an assemblage or as a cause to an effect; instead, so-called agency becomes an epiphenomenon associated with observations and perceived effectiveness.

Agency or Efficacy

After reviewing several schools of thought on the notion of nonhuman agency, a preliminary conclusion may be made to clarify what I mean by nonhuman agency, or agency in general, in a posthumanist framework. Two models have emerged for conceptualizing nonhuman agency. The first is the notion of "distributed agency" or "assemblage agency" as exemplified by certain versions of ANT and new materialism, represented by Jane Bennett's vital materialism. The second is "agential agency," represented by Levi Bryant's machine-media ontology and systems theory. These seemingly conflicting approaches reveal a tension that reminds us of two crucial aspects when considering agency in a posthumanist framework.

First, assemblage agency is a reminder that perceived effectiveness involves more than an agent. The individuation process fixes a scale for observation. It uses observed effectiveness to generalize a field of heterogeneous components, as if the perceived individual is able to act as the sole source of agency and as a cause to an effect. Individuation overlooks that perceived agency is historically contingent, and the components might have been and can be reconfigured in different ways. For example, we attribute observed effectiveness in nature and technology as two generalizing terms, as if nature and technology are the ultimate causes that initiate change. However, as we saw in Part One, these terms have been constructed as arrangements of knowledge, and might have been arranged differently. Individuation makes salient a "machine" by dimming the "media" around it. An agent does not exist within a vacuum, and its perceived effectiveness is always achieved through other assemblages functioning as media. Another example is the extended mind theory, which posits that humans devolve cognitive functions to the environment. The human image localizes agency within a human body that is, in fact, deeply embedded in a network of systems that make us appear to initiate changes.

Second, agency can be understood as "perceived agency" or "observed effectiveness." When Neil Leach critiques Heidegger for ascribing agency to an object, he argues that it is a form of "ventriloquism, of projecting onto the object a form of anthropomorphic agency."[50] Ventriloquism? True. Yet, I argue that Leach is not stretching the metaphor far enough. From a posthumanist perspective, what Leach overlooks is that not only is ascribing agency to an object a form of ventriloquism, but also ascribing agency to the human is, as well, in that it constructs an image of the human as a source from which actions are generated. In this way, Bryant's agential agency reminds us that ascribing agency, human or nonhuman, exists in concert with a conceptualization of an agent. Thus, I propose a compatibility thesis between agency and individuation: *agency cannot be separated from the individuation process, and both result from observation*. The treatment for agency as an observational strategy in a field of heterogeneous assemblages renders agency an epiphenomenon, an effect rather than a cause; it is not because an assemblage exhibits a certain degree of agency that it can be counted as an agent, but because we individualize an assemblage as an agent that it then becomes a carrier of distributed agency.

Most importantly, observations of nonhuman agency are made by humans, so what we understand as nonhuman agency "becomes a matter of attribution, *post hoc* and after the action." Through observation, "some entities are detached from their background and called 'actors'. They are made to conceal and stand for the web of relations that they cover. They become the place where explanation, moral, causal, practical, stops."[51]

Searching for "nonhuman agency" thus becomes a false quest; in the end, it concerns summoning more nonhuman agents to expand the field of perceived effectiveness around humans. From this vantage, many posthumanist ideas, at their best, remove the human from the center of agency only by presenting a field of heterogeneous assemblages, or a network of actors where agency is distributed. However, any attempt to describe what nonhuman agency is, such as the transformative agency of trees, points to efficacy, which forces nonhumans to act in a tale which we tell ourselves to diversify and hybridize human agency. This predicament is why Harman attempts to distinguish OOO from posthumanism, and argues that posthumanism has alluded to a new paradigm but fails to provide what a truly posthumanist framework might look like; toward that end, OOO attempts to provide such a posthumanist framework.[52] This is, in part, the reason why OOO is regarded by many as controversial, because OOO is essentially investigating a different paradigm. When OOO asserts that objects are withdrawn and thus "do not act," many scholars would interpret this claim as a revolt against ANT, overlooking its true message. If we respect this posthumanist paradigm shift, OOO simply does not care if an object acts or not; instead, it asks about the possibilities of an object beyond *acting*.

Notes

1 Bennett, *Vibrant Matter: A Political Ecology of Things*, 229.
2 Bennett.
3 Bryant, *Onto-Cartography: An Ontology of Machines and Media*.
4 Harman, *Object-Oriented Ontology: A New Theory of Everything*.
5 Bijker, "How Is Technology Made?—That Is the Question!"

6 Müller and Schurr, "Assemblage Thinking and Actor-Network Theory: Conjunctions, Disjunctions, Cross-Fertilisations," 220
7 JonesandCloke,*Tree Cultures: The Place of Trees and Trees in Their Place*; "Non-Human Agencies: Trees in Place and Time," 79–96.
8 Jones and Cloke, "Non-Human Agencies: Trees in Place and Time," 81.
9 Jones and Cloke, 81.
10 Jones and Cloke, 84.
11 Jones and Cloke, 85.
12 Jones and Cloke, 80.
13 Jones and Cloke, 85.
14 Müller and Schurr, "Assemblage Thinking and Actor-Network Theory," 218.
15 Wolfe, *What Is Posthumanism?*, 125.
16 Müller and Schurr, "Assemblage Thinking and Actor-Network Theory," 226.
17 Müller and Schurr, 217.
18 DeLanda, *Assemblage Theory*.
19 Müller and Schurr, "Assemblage Thinking and Actor-Network Theory," 220.
20 Bennett, *Vibrant Matter: A Political Ecology of Things*, 23.
21 Bennett, 31.
22 Bennett, 35.
23 Bennett, 113.
24 Bennett, 112.
25 Brett Milligan, "Vibrant Matter and Relations of Things," *Free Association Design* (blog), September 26, 2011, https://freeassociationdesign.wordpress.com/2011/09/26/vibrant-matter-and-relations-of-things/. Last accessed: April 29, 2024.
26 Bennett, *Vibrant Matter: A Political Ecology of Things*, 151.
27 Harman,"Object-Oriented Ontology."
28 Harman.
29 Harman, *Object-Oriented Ontology: A New Theory of Everything*.
30 Harman.
31 Harman.
32 Leach, "Digital Tool Thinking: Object-Oriented Ontology versus New Materialism."
33 Leach, 346.
34 Leach.
35 Bennett, "Systems and Things: On Vital Materialism and Object- Oriented Philosophy," 17.
36 Harman, *Object-Oriented Ontology: A New Theory of Everything*, 242.
37 Harman, 242.
38 Harman, 163.
39 Harman, 189.
40 Harman, 189.
41 Harman, 241.
42 Bryant, *Onto-Cartography: An Ontology of Machines and Media*, 286.
43 Bryant, 193.
44 Bryant, 19.
45 Bryant, 199.
46 Bryant, 200.
47 Bryant, 201.
48 Bryant, 219–20.
49 Bryant, 223.
50 Leach, "Digital Tool Thinking: Object-Oriented Ontology versus New Materialism," 348.
51 Law and Mol, "Non-Human Agencies," 58.
52 Harman, "Object-Oriented Ontology."

Bibliography

Bennett, Jane. "Systems and Things: On Vital Materialism and Object- Oriented Philosophy." In *The Nonhuman Turn*, edited by Richard Grusin, 17. Minneapolis; London: University of Minnesota Press, 2015.

———. *Vibrant Matter: A Political Ecology of Things*. Durham, NC: Duke University Press, 2010.

Bijker, Wiebe E. "How Is Technology Made?—That Is the Question!" *Cambridge Journal of Economics* 34, no. 1 (January 1, 2010): 63–76. https://doi.org/10.1093/cje/bep068.

Bryant, Levi R. *Onto-Cartography: An Ontology of Machines and Media*. Speculative Realism. Edinburgh: Edinburgh University Press, 2014.

DeLanda, Manuel. *Assemblage Theory*. Speculative Realism. Edinburgh: Edinburgh University Press, 2016.

Harman, Graham. "Object-Oriented Ontology." In *The Palgrave Handbook of Posthumanism in Film and Television*, 401–9. London: Palgrave Macmillan, 2015. https://doi.org/10.1057/9781137430328_40.

———. *Object-Oriented Ontology: A New Theory of Everything*. London: Penguin, 2018.

Jones, Owain, and Paul Cloke. "Non-Human Agencies: Trees in Place and Time." In *Material Agency: Towards a Non-Anthropocentric Approach*, edited by Carl Knappett and Lambros Malafouris, 79–96. Boston, MA: Springer US, 2008. https://doi.org/10.1007/978-0-387-74711-8_5.

———. *Tree Cultures: The Place of Trees and Trees in Their Place*. Oxford; New York: Berg, 2002.

Law, John, and Annemarie Mol. "The Actor-Enacted: Cumbrian Sheep in 2001." In *Material Agency: Towards a Non-Anthropocentric Approach*, edited by Carl Knappett and Lambros Malafouris, 57–75. Boston, MA: Springer US, 2008. https://doi.org/10.1007/978-0-387-74711-8_5.

Leach, Neil. "Digital Tool Thinking: Object-Oriented Ontology versus New Materialism." In *Proceedings of the 36th Annual Conference of the Association for Computer Aided Design in Architecture*, 344–51. Ann Arbor, MI: ACADIA, 2016.

Milligan, Brett. "Vibrant Matter and Relations of Things." *Free Association Design* (blog), September 26, 2011. https://freeassociationdesign.wordpress.com/2011/09/26/vibrant-matter-and-relations-of-things/.

Müller, Martin, and Carolin Schurr. "Assemblage Thinking and Actor-Network Theory: Conjunctions, Disjunctions, Cross-Fertilisations." *Transactions of the Institute of British Geographers* 41, no. 3 (July 1, 2016): 217–29. https://doi.org/10.1111/tran.12117.

Wolfe, Cary. *What Is Posthumanism?* Posthumanities Series. Minneapolis: University of Minnesota Press, 2010.

7

FROM NONHUMAN AGENCY TO SPECULATIVE ONTOLOGY

Posthumanism entails two levels of reflexivity concerning agency. The first level of reflexivity may be described as a nonhuman turn, a shift from the social constructivism prevalent throughout the 1980s to a focus on the nonhuman realm and its agency in shaping humanity toward the early 2000s. In the second half of the twentieth century, the social construction of nature and social construction of technology (SCOT) ascended as two fields of research offering critical perspectives on the conception of nature and technology. Although they came from different traditions with distinct frameworks, a common theme emerged in both fields—the critique of human exceptionalism. Namely, actor-network theory (ANT) ascended from SCOT tradition and became a more general mode of thinking across fields; ANT offered an analytical framework to consider nonhuman actors and their role in mobilizing social groups and forming actor networks where human agency resides. The notion of distributive agency became a common theme in the 2010s.

However, as many posthumanist scholars have struggled to comprehend, even though we embrace agency's distributive quality, we must daily confront individual objects and view them as actors and agents. Thus, one theme cutting across posthumanist ideas is the tension between the distributive quality of agency and the stubborn reality of the individuation of objects as carriers of observed effectiveness. For instance, "actor-network" is an oxymoron; we must accept a thing as an actor and, simultaneously, a network.

The second level of reflexivity is represented by a backlash in searching for nonhuman agency. The reflection of the irony of nonhuman agency has led to the advancement of speculative realism and object-oriented ontology (OOO) since the 2010s. Because stories of how nonhumans structure the human world are inevitably told through human narrators, nonhuman

DOI: 10.4324/9781003320852-10

agency becomes a *post hoc* attribution of observed effectiveness, based on human standards. Applying "agency" as a framework, which is, after all, a human-centric concept, to analyze the nonhuman realm ironically advanced another level of human hubris. Hence, many scholars have struggled with confusion and ambiguity regarding the term *agency*, with respect to identifying wherein lie cause and effect.

The incentive to focus on nonhuman agency was to challenge human-centric views by searching for what enables nonhuman actors to act. This is partly why assemblage thinking and new materialism have become increasingly popular since the late aughts. Scholars such as Jane Bennett have built frameworks upon material agency, imbuing their investigations with mystical terms such as "vibrant matter" or "matter-energy," which allow bodies to interact with each other and form assemblages to gain power.[1] Yet, no matter how many nonhuman agents we summon to diversify our narratives, we end with stories of observed effectiveness relevant only to humans. Nonhumans are always actants following the plot of the storyteller. We may conceptualize agency in multiple ways—distributed or localized, human or nonhuman, material or immaterial—but, in sum, the term *agency* cannot capture an entity's complete reality.

Once a framework begins to produce paradoxes rather than transformative thinking, it is time to reflect on the framework itself. Many posthumanist scholars have recognized this irony, and accepted the existence of "surplus" in both human and nonhuman assemblages that are irreducible to any form of (human) access. However, because these ideas mainly rely on agency as an important component in their conceptual framework, they are haunted by the inherent paradox.

Since the early 2010s, OOO has gained popularity across fields, especially in architecture and landscape architecture. Here, we must distinguish OOO as a conceptual framework operating in a different paradigm from other posthumanist ideas, such as assemblage thinking and new materialism (Figure 7.1). They resemble each other in terms of what they critique: the unquestioned anthropocentrism at every level of our thinking since the Enlightenment. Yet OOO takes a different approach, bypassing the notion of agency—a human concept—and directly querying the "surplus" in objects, speculating as to what the world might be like in itself. Hence, OOO is

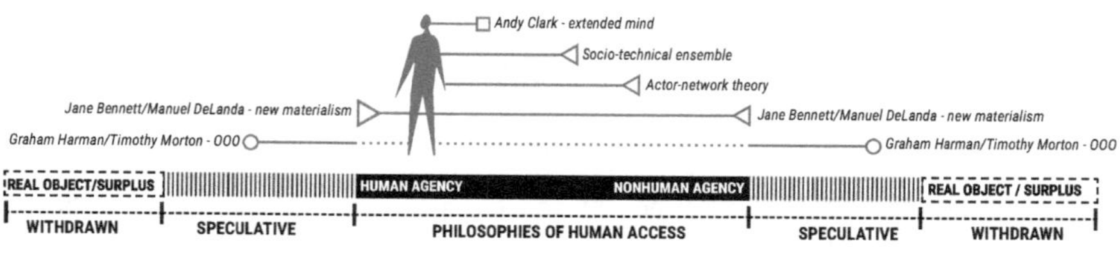

FIGURE 7.1 Posthumanist speculations. Illustration by the author.

often considered a major tenet in the alleged "speculative realist movement" despite internal debate and criticism about whether such a movement exists.

At this point, we must recognize that the genealogy presented in this book joins with a broader philosophical reflection across fields. I provide only a limited perspective concerning nature, technology, and intelligent machines in this broad-based effort to reflect on our human ways of thinking, or what Graham Harman would call "philosophies of (human) access." It reflects the "prevalent tendency within Kantian and post-Kantian thought to treat the relation between thought and world as the primary subject matter of philosophy."[2] On this path of reflection, many, such as Bennett, provide ethical and political concerns; others, such as ANT scholars, offer concerns that are more epistemological. In the end, it is more aptly an ontological concern since multiple tenets of posthumanism are, to use Harman's terms, "philosophies of access" limited by their "inability—or better, unwillingness—to create a speculative ontology which moves beyond the narrow confines of what is given to our all-too-human modes of understanding."[3]

Furthermore, we need to acknowledge that a further philosophical project which OOO seeks to achieve is a reflection on not only human access but also on all types of access between assemblages. Posthumanist tenets such as ANT, assemblage thinking, and new materialism are all relational, based on the premise that things can interact, communicate, form relational bonds, and enable each other. The relations between them are *a priori* for objects to achieve significance. This is best exemplified by Bennett's claim that, "bodies enhance their power in or as a heterogeneous assemblage."[4] Indeed, this form of relational thinking aligns comfortably with modern society's ecological concerns, also often understood as relational and active. Moreover, a positive ontology is no doubt favored by the mainstream ideology of diversity that encourages communication and interaction between and among differences.

However, this relational way of thinking runs the risk of undermining and overlooking an entire spectrum of thinking based on the assumption that things do *not* interact, communicate, and relate to each other. It is not that they have no desire to do so, but they are unable to interact and communicate in the first place. This assumption is an expansion of Niklas Luhmann's (in)famous claim that "humans cannot communicate."[5] This is where OOO differentiates itself from other posthumanist thinking: it seeks to create a non-relational ontology. Here, we need to further recognize why "agency" is irrelevant in OOO's framework—that is, to believe that our thought (or anything's thought) forms a relationship with the world and other thoughts is *a priori* to conceptualizing agency. To think we can communicate provides the initial urge and motivation to act and to control. As we shall see in later chapters, this mode of reasoning has also manifested in cybernetics and modern control theory, which are based on the premise that recursive processes lead to communication, and communication leads to control.

Thus, OOO seeks to render a speculative ontology so that we may begin to consider reality in a non-relational way and speculate how things possibly "interact." On the path toward a speculative ontology, two major approaches exist within OOO's framework. The first approach is to consider what it is to perceive efficacy from a nonhuman perspective. This is known as "alien phenomenology," proposed by OOO scholar Ian Bogost and widely endorsed by OOO proponents. The second approach is to construct metaphors, as proposed by Graham Harman. Building on these two approaches, I will argue that "design" is a third approach to speculative ontology.

Alien Phenomenology

In truth, OOO scholars are not the first to consider what it is to be another thing. In 1974, American philosopher Thomas Nagel proposed a question in a paper titled "What Is It Like to Be a Bat?". Nagel argued that since we know a bat navigates through echolocation and other behavioral aspects, we may imagine what it is to *behave* like a bat by taking on a bat's point of view; but we will never know what it is *for* a bat to be a bat. His widely cited assertion held that, "an organism has conscious mental states if and only if there is something that it is like to be that organism—something it is like *for* the organism."[6]

However, when we take a posthumanist twist of Nagel's claim, not only are we unable to access the bat's experiences as a bat, but it also appears impossible to understand completely what it is to be human. Assemblage framework frames what we identify as human consciousness and experience as epiphenomena, and emergent properties of chaotic interactions between distributed systems within and around the human body as the localized carrier of agency; thus, it is impossible to define all the systems involved in making us humans.

Inaccessibility is the starting point of alien phenomenology. If accessing both human and nonhuman experiences becomes the same practice of inferences, then what are the analytic, ethical, and political implications for inferring what is to be anything other than human? By asking this question, alien phenomenology promotes a mode of analysis with a sense of extra attentiveness toward assemblages rather than individuals, in terms of how they interact with and co-produce each other. Ethically, it "opens the possibility of more compassionate ways of relating to aliens, helping us to better attend to their needs, thereby creating the possibility of better ways of living together."[7] Politically, alien phenomenology may help us understand the flows and operations of large-scale assemblages such as corporations, and "increases the efficacy of our political intervention."[8]

Alien phenomenology turns Nagel's argument, which questions how one can verify what it is to think like a nonhuman assemblage, into a focus on the implications of such considerations. Even though ascribing agency to nonhumans is a false premise, since it ends up being efficacy and perceived effectiveness, the way in which we ascribe agency retains profound ethical and moral implications. The issue at hand becomes one of which type of conceptualization of efficacy will best serve the needs of humanity, particularly when facing challenges such as the climate crisis. Toward this end, nearly all posthumanist ideas arrive at versions of political ecology that promote the belief that the future ought to be co-produced by more-than-human assemblages. From this vantage, alien phenomenology is never about aliens but about ourselves as humans. It is a sort of thought experiment that requires more doing than proving, which helps us stretch the sphere of our moral and ethical considerations to include more of those entities which we thought of as "other" and to cultivate a sense of care and empathy.

One can consider landscape design as a practice of alien phenomenology. When speaking of designers approaching landscape, some landscape architects aver that they must think like mountains, like rivers, and like trees.[9] To a certain extent, landscape architects are experts in alien phenomenology, by considering how different human and nonhuman assemblages interact with their environment and each other, as well as how these interactions may be transformed into creative strategies that foster a shared environment.

Metaphor

When we ascribe agency, what becomes invisible is the "surplus," "real object," or "dark object," a common theme emergent in most of the posthumanist ideas, except for certain versions of ANT. They all posit the existence of something beyond any form of access; in many cases, the terms "nonhuman agency" and "material agency" were employed. Ironically, the moment "agency" is used to denote some newly discovered efficacy of an assemblage—for example, the transformative agency of trees in place-making—the term ceases to denote "surplus" and instead speaks to specific relationships that define the assemblage as an actor. To use Harman's term, ascribing agency thus becomes a form of overmining, which forces the assemblage to act in an actor-network mobilized by humans.

Since agency is understood in conjunction with individuation to denote perceived effectiveness, we may detach the notion of agency from the concept of "surplus" or "dark object." In fact, Harman seldom uses the term "agency" in his version of OOO, which is not interested in how objects act, especially for humans. Instead, OOO seeks to query the "surplus" in objects, completely bypassing the concept of agency.

The theme of "surplus" accompanies a predicament. It is a tension between the stubborn reality of ubiquitous nonhuman agency, and our urge to seek and embrace the inaccessible surplus in assemblages. In other words, observing effectiveness and ascribing agency in assemblages within a posthumanist framework becomes a simple task; seeing surplus is difficult. The real surplus we seek is overwhelmed by the omnipresence of agency. The problem of agency is that it veils an assemblage's other potentials, including the potential along its own trajectory, or intrinsic values beyond any form of human frameworks.

The inaccessibility of the real object makes Harman reject "scientism" and "mathematism" and turn to art and aesthetics. Harman believes "metaphor" provides a form of non-literal access to real objects. He uses Homer's "wine-dark sea" as an example to illustrate how a metaphor works in OOO's analytical framework. In this metaphor, the unusual, sensual quality of "wine-dark" no longer belongs to the sensual object of "sea"; it points to the real object of "sea"—the unrealized surplus. However, a real object is withdrawn from access, and the real object of "sea" cannot be accessed. In Harman's framework, quality must be paired with an object, so there must be a real object to embrace this unusual, sensual quality of "sea" being "wine-dark." In this case, the aesthetic beholder, the only real object in this object-quality pairing, will function as a real object to retain the metaphor. Consequently, Harman argues that aesthetic experiences such as metaphors are highly valued in OOO's framework.

To further illustrate Harman's point, we shall unpack this by turning to Harman's critique of scientism. Sciences provide us with tools with which to access objects by overmining or undermining them—to study what a thing is made of or how it can be used. However, an object should not be solely defined by sciences, and there should be a surplus of an object which sciences fail to provide with tools to access. Metaphors thus become instruments for mining this "third object" beyond a thing's material components and its use. Inventing metaphors becomes a way to break away from scientism's "dual-mining" that forces objects to be effective and useful in human discourse.

Design

As discussed above, OOO has gained increased attention within the design professions, especially in architecture, because it provides a framework to re-focus on architectures as objects with qualities. Toward the late twentieth century, the rise of critiques on everyday life, urban geography, and urban studies transformed buildings into a byproduct of socio-cultural systems. In response to these emerging critiques, many in the discipline began to adopt systems thinking to approach architectural design, and buildings were envisaged through a set of social, economic, ecological, and urban systems. Since the 2010s, certain architecture theorists have begun to resort to OOO due to "a frustration that architecture is increasingly justified solely by its relations and not by its own particular and autonomous qualities."[10] In this wave of "object-turn," the Southern California Institute of Architecture (SCI-Arc) has become a pioneer. Graham Harman has taught philosophy at SCI-Arc since 2016; in 2019, another key figure in the OOO movement, Timothy Morton, joined him as a visiting faculty member in the synthetic landscape program. Thus, OOO has directly influenced SCI-Arc students' architectural design projects over the past several years. These projects focus on the architectural objects themselves rather than on the relationships in which they are enrolled.

In contrast to architects' approach to architectural design with OOO, the landscape architecture discipline has formed a different perspective on OOO, thanks to Timothy Morton. Morton conceptualized the idea of *hyperobject* "to refer to things that are massively distributed in time and space relative to humans,"[11] such as climate change. By framing climate change as a *hyperobject*, Morton makes it inaccessible; local manifestations of climate change, such as sea level rise, storm surges, heat waves, and extreme drought, are not climate change itself. Hence, the challenge is indirectly accessing climate change as an object through design. Indeed, current discussion reduces "climate change" into the perceptible and measurable effects it causes, such as flooding and extreme heat. According to David Ruy, the Postgraduate Program Chair at SCI-Arc, the focus of the synthetic landscape program is to address the issue of climate change, which "is an existential drama that is now unfolding at a planetary scale."

Furthermore, Ruy has referred to Timothy Morton's *Ecology Without Nature*. In this book, Morton argues that nature is a useless concept in ecological thinking, because ecological thinking "has set up 'Nature' as a reified thing in the distance, under the sidewalk, on the other side where the grass is always greener, preferably in the mountains, in the wild."[12] Morton's argument echoes what Ruy believes:

> ... in contrast to traditional landscape design programs that emphasize the stewardship of nature and traditional western cultural values of picturesque or sublime images of nature, this program will challenge the status quo. Instead of being trapped in an environmental melancholy, we would like to see if our design imagination can project an abundant future with new forms of beauty.[13]

I empathize with SCI-Arc's environmental imperative; however, the SCI-Arc method of landscape architecture is not without precedents, and these ideas are hardly novel within the "traditional" landscape design discipline. For example, landscape theorist Elizabeth Meyer has discussed hybridity and cyborg landscapes since the 1990s, and they were well-received in the discipline by the 2000s. In a way, landscapes are "synthetic" by definition since they result from the co-production of heterogeneous assemblages, including humans, nonhuman species, and machines. Landscape architecture is a discipline that grew as a "modern other" and has been theoretically and historically misinterpreted.[14] The urgency here is not in how to "reinvent" landscape and redefine beauty, but in how to challenge the stereotypical understanding of landscape as picturesque, which is how it is widely accepted by the general public. Toward this end, it has already been argued that the task for today's landscape architects is to cultivate new categories of aesthetic experience, beyond beautiful, sublime, and picturesque; numerous designed projects in this vein have been identified and discussed.[15]

As a design institute, SCI-Arc has provided an interpretation of OOO in the design professions of landscape and architecture. This effort should be acknowledged and supported because to create this institutional initiative requires the courage to admit, confront, and challenge the marginalized position of "design" in contemporary culture. In a way, SCI-Arc's efforts may be viewed as a shared frustration within the design professions in response to the irony that the more "design thinking" is talked about outside "design disciplines," the less it is about "design." In mainstream, techno-scientific knowledge production, design as a practice that produces another category of knowledge has been either ignored or misinterpreted as "creative problem-solving." In response to the marginalized position of design, OOO could offer a new framing for "design."

What is valuable about metaphor is an unusual connection between two objects; from this perspective, design continually creates new metaphors that fashion distinctive connections between objects. On that note, OOO architects' re-focus on the architectural object must not be a conclusion, but rather a starting point for inquiring what new sort of unusual relations may be cultivated between a building and its environment, beyond the received model that defines buildings as components in socio-cultural systems. Similarly, the synthetic landscape is less about reinventing "machines in the garden"—technical objects interacting with natural objects—than asking what other new relationships may be observed in the field of heterogeneous assemblages of machines, humans, and nonhuman species. Mapping new relationships is what landscape architects have done for decades.

For example, SCAPE's oyster-tecture project builds an unusual connection between two objects—public space and oysters. The proposal reveals a new aspect of the "surplus" in oysters, as one co-producer of a new public space. Landscape design allows oysters to manifest their effectiveness beyond their definition as oysters. Yet this unusual connection also reveals what the object of "public space" might mean for a more-than-human public. A landscape then becomes a new front from which to explore novel relationships between machines, humans, and other assemblages in the environment, and how they might co-produce a future.

However, once a metaphor is made and widely accepted, it ceases to be a metaphor; it is instead a reified relationship in our daily encounters. To a certain extent, relations with which we are familiar today were once, in the past, metaphors. Indeed, the tale of oysters and the public landscape is no longer novel, since SCAPE's oyster project has been discussed over the years; thus, oysters' effectiveness in cleaning water and participating in place-making has become a reified agency in the oyster. It now appears nearly trendy to propose a version of oyster infrastructure for a coastal area. This *post hoc* attribution of agency suppresses the oyster's potential effectiveness in other metaphors and the potential for being simply a sea creature. Design becomes an act of continually searching for new metaphors between objects, an act of exploiting the inexhaustible "surplus" from which effectiveness emanates. For this reason, in design professions, designers tend to avoid analyzing "problems"; rather, they explore "potentials." Replacing "problems" with "potentials" is more than wordplay; it is a constant reminder of the inexhaustible surplus in the most mundane of daily assemblages.

Therefore, design may be understood as an act that identifies unusual connections in the field of heterogeneous assemblages; it is a form of speculation that provides indirect access to withdrawn objects.

How might design create new metaphors between assemblages? How would different assemblages interact with each other, and how do they reveal and exploit each other's effectiveness? These questions paraphrase another: how do assemblages, human or nonhuman, living or non-living, biotic or abiotic, biological or technical, co-produce each other and the shared environment? Metaphors and design involve creativity and originality, qualities that indicate intelligence rather than agency. There is always that within an assemblage that allows it to reveal and exploit effectiveness in other assemblages. This sort of ability may be understood through intelligence, the ability to construct knowledge and apply it so that effectiveness emerges. In the next chapter, I will explore the concept of intelligence, considering the rapidly evolving field of artificial intelligence within a posthumanist sensibility.

Notes

1 Bennett, *Vibrant Matter: A Political Ecology of Things.*
2 Young, "On Correlationism and the Philosophy of (Human) Access," 43.
3 Young, 51.
4 Bennett, *Vibrant Matter: A Political Ecology of Things*, 23.
5 Luhmann, "How Can the Mind Participate in Communication?"
6 Nagel, "What Is It Like to Be a Bat?," 436.
7 Bryant, *Onto-Cartography: An Ontology of Machines and Media*, 70.
8 Bryant, 71.
9 I owe landscape architect Teresa Gali-Izard credit for this satement. She used such sayings in her lectures and daily conversations when I was a landscape architecture student of hers.

10 Gage, "Killing Simplicity," 95.
11 Morton, *Hyperobjects*, 1.
12 Morton, *The Ecological Thought*, 3.
13 SCI-Arc, "SCI-Arc Launches New Postgraduate Program, Synthetic Landscapes – SCI-Arc."
14 Meyer, "The Expanded Field of Landscape Architecture."
15 Meyer, "Sustaining Beauty. The Performance of Appearance"; Meyer, "Values in Landscape Architecture."

Bibliography

Bennett, Jane. *Vibrant Matter: A Political Ecology of Things*. Durham, NC: Duke University Press, 2010.

Bryant, Levi R. *Onto-Cartography: An Ontology of Machines and Media*. Speculative Realism. Edinburgh: Edinburgh University Press, 2014.

Gage, Mark Foster. "Killing Simplicity: Object-Oriented Philosophy In Architecture." *Log*, no. 33 (2015): 95–106.

Luhmann, Niklas. "How Can the Mind Participate in Communication?" In *Theories of Distinction*, translated by William Rasch, 169–86. Redescribing the Descriptions of Modernity, Cultural Memory in the Present. Stanford, CA: Stanford University Press, 2002.

Meyer, Elizabeth K. "Beyond 'Sustaining Beauty.'" Edited by Ann Komara. *Values in Landscape Architecture: Finding Center in Theory and Practice* 71, no. 1 (January 2, 2017): 125–27. https://doi.org/10.1080/10464883.2017.1260964.

———. "Sustaining Beauty. The Performance of Appearance: A Manifesto in Three Parts." *Journal of Landscape Architecture* 3, no. 1 (March 2008): 6–23. https://doi.org/10.1080/18626033.2008.9723392.

———. "The Expanded Field of Landscape Architecture." In *Ecological Design and Planning*, edited by George F. Thompson and Frederick R. Steiner, 45–79. New York: John Wiley & Sons, Inc., 1997.

Morton, Timothy. *Hyperobjects: Philosophy and Ecology after the End of the World*, 1st ed. Minneapolis: Univ Of Minnesota Press, 2013.

———. *The Ecological Thought*. Reprint edition. Cambridge, MA: Harvard University Press, 2012.

Nagel, Thomas. "What Is It Like to Be a Bat?" *The Philosophical Review* 83, no. 4 (October 1974): 435. https://doi.org/10.2307/2183914.

SCI-Arc. "SCI-Arc Launches New Postgraduate Program, Synthetic Landscapes - SCI-Arc," April 2009. https://www.sciarc.edu/news/2019/sci-arc-launches-new-postgraduate-program-synthetic-landscapes.

Young, Niki. "On Correlationism and the Philosophy of (Human) Access: Meillassoux and Harman." *Open Philosophy* 3, no. 1 (January 1, 2020): 42–52. https://doi.org/10.1515/opphil-2020-0003.

8

CO-PRODUCTIVE INTELLIGENCE

Although posthumanist scholars have given sufficient account of nonhuman objects and assemblages, occasionally including infrastructures and computer software, their investigations leave the concept of intelligence unexplored. Many posthumanist ideas have focused solely on agency, not on the intelligence intrinsic to originating a sense of agency; we assume an object that acts is intelligent. Recent AI and machine-learning developments have provided empirical material to consider intelligence with a posthumanist reframing.

Over the past decade, AI has become pervasive in contemporary culture. Media coverage on AlphaGO and its predecessors, and other AI systems, led to further rounds of excitement and fear for the possibilities of artificially intelligent machines. Since mid-2022, generative AI systems like ChatGPT and image generators like Midjourney and Stable Diffusion have flooded social media; AI is no longer a farfetched concept from science fiction but applications in your browsers. This development propels one to speculate how to integrate AI in environmental systems, and poses a question: Will the exponential growth of AI applications shift the role of machines—other than as tools of optimization and control—in how the environment is conceived and constructed?

To answer, we must understand not only the technical capacity but also the mode of thinking that undergirds mainstream AI research. I will analyze cases in AI research with a posthumanist lens to evoke their inherent presumptions in understanding machines as tools of optimization that extend imagined human agency in managing the environment. Even though the boundary between nature and technology dissolves in the field of heterogeneous assemblages (see Part I), agency's individuation process leads to tenacious objects such as machines, plants, animals, and humans. To consider

DOI: 10.4324/9781003320852-11

intelligent machines in the environment risks falling into the normalized "adaptive management" that uses advanced technologies to automate and optimize natural processes; machines are regarded as a layer of control strategies through which humans manage natural processes.

Instead, I argue that intelligence cuts across different assemblages and reveals shared abilities or tendencies of the assemblages to exploit each other's surplus and produce effects. Most importantly, unlike agency that emphasizes individuation, intelligence involves interactions between individuals. However, this argument faces three challenges: *anthropocentrism*, *intelligence as a measurement*, and *individualism*. By deconstructing these three obstacles, I will restructure a posthumanist concept of co-productive intelligence.

Anthropocentrism and Universal Intelligence

The Merriam-Webster online dictionary defines intelligence as

> the ability to learn or understand or to deal with new or trying situations [...] the ability to apply knowledge to manipulate one's environment or to think abstractly as measured by objective criteria (such as tests) [...] the act of understanding.[1]

Though unspecified, these definitions point to human cognitive capacity, including learning and applying knowledge. In other words, to discuss and define intelligence, we maintain a human image in our minds and then venture to consider cognition in nonhuman entities, such as animals and machines. Anthropocentrism and human exceptionalism are intrinsic in the conceptualization of intelligence. However, a posthumanist framework is interested in how different assemblages, human and nonhuman, living and nonliving, coproduce each other, and this presents anthropocentrism and human exceptionalism with the first obstacles to overcome. In recent years, the research on machine intelligence has shed light on this issue and may help to map a posthumanist understanding of intelligence.

Machine intelligence has long been an important frontier in intelligence research. Can machines think? In 1950, Alan Turing proposed this question and the famous Turing test to argue that machines *can* think. The Turing test involves three players: a human discriminator in one room interviewing a human player and a machine in a separate room. The discriminator will try to discern between the machine and the human by asking questions without seeing them. The machine will try to imitate the human player to deceive the discriminator into believing that the machine is a human. Another concept Turing proposed was "child-machine," a theoretical strategy for machines to learn through random mutations and natural selection.[2] This strategy underpins many of today's most advanced machine-learning techniques.

The Turing test and the "child-machine" are evidence that, from its earliest stages, artificial or machine intelligence has been envisioned based on human intelligence. The Turing test essentially proposes to evaluate machine intelligence based on its resemblance to human behavior. We have long compared machines to humans; the first thing people do to test a powerful machine is to match it with the best of its human competitors. Over the years, many cases have been reported and discussed: Deep Blue (1997) defeated Garry Kasparov, the best human chess player; AlphaGo and its successors like AlphaZero and MuZero (2015–2019) trounced many of the best GO players and mastered Atari video games; AlphaStar (2019) mastered StarCraft, a real-time strategy video game notorious for decision-making based on incomplete knowledge; and OpenAI Five (2016–2019), a team of five separate AI agents, worked together and outperformed a human team in the game of DOTA2, known for its real-time collaborative strategies and corporations, as well as the ability to understand another player's intentions and act accordingly.

Underpinning our obsession with pitting humans against machines is, ironically, a sense of human exceptionalism—the unstated belief that humans are the most intelligent entities on Earth. If a machine outperforms, or at least matches, humans in one aspect, that is seen as a breakthrough. It is an ambivalent state of mind. The notion that another entity could replace humans at the top of the intelligence pyramid evokes terror, yet we secretly enjoy this sense of unease through the reassurance that we are the ultimate creators of intelligent machines. Our obsession with human-machine intelligence reveals another level of technological sublimity—the mixed sense of terror and joy. This state of mind was fruitful for popular culture products over past decades, making technological dystopia popular in video games, films, and television. Many televised franchises imagined a version of rogue AI attempting to end humanity. However, accompanying our terror is a sense of sublimity because these are, after all, fantasies that pose no real threat. Most importantly, these shows inevitably end with humanity regaining control over the rogue AI, reassuring the audience with a sense of human exceptionalism.

Other forms of intelligence are also measured against human intelligence and cognitive abilities. For example, when evaluating animal cognition, the common criteria used include teaching, short-term memory, causal reasoning, planning, deception, transitive inference, theory of mind, and language. Based on these criteria, a significant gap has been conceptualized between human and animal cognition.[3] Nevertheless, these evaluative criteria are modeled in favor of the human. Anecdotes often compare dogs to two-year-old human toddlers and dolphins to three-year-olds. Yet, this type of comparison essentially treats animals as diminished versions of us, reflecting deep human exceptionalism in framing the concept of intelligence. If there were a test of intelligence based on olfactory senses, many animals, such as dogs, would be far more intelligent than humans. Moreover, plant intelligence has emerged as a topic. In one study, researchers demonstrated that plants possess learning and memory, too, through systems unique to them. For example, *plasmodesmata*, a type of intercellular organelle found only in plant and algal cells, are crucial for plants to transmit information.[4] Yet research still attempts to locate human cognitive faculties in plants rather than discussing plant intelligence *for* plants. These examples speak to an intrinsic bias in our examination of nonhuman intelligence. By comparing other entities with the human image, we deny the possibility of noting and embracing how nonhuman entities interact with their environments in entirely different ways.

Furthermore, speaking of human intelligence unwittingly evokes an image of a perfect human being. This gives rise to a series of problematic interpretations similar to those that fueled scientific racism and discrimination narratives through the twentieth century. The concept of intelligence has long been intertwined with the intelligence quotient (IQ) test. French psychologist Alfred Binet invented the first practical IQ test, which was later translated into English and revised in 1916 as the Stanford–Binet Intelligence Scales. Intelligence tests were then quickly adopted as tools to explain racial, class, and sex differences—however narrowly "intelligence" was defined by the tests—justifying all sorts of disturbing narratives of colonialism, slavery, social Darwinism, sexism, and racist eugenics.[5]

The problem of intelligence is not the concept itself, but using a white European male image of humanity as a reference for measurement. It ignores the fact that our cognitive functions result from the long-term co-evolution of human and other nonhuman assemblages, including other species, languages, tools, and other distributive systems.

Taking human-machine intelligence as an example, we will see that what we thought was human intelligence may instead be reframed as an outcome of co-production between humans, transistors, and circuit boards. The central processing unit (CPU) in any computer is a conglomeration of transistors wired in specific ways to compute logical and arithmetic statements through the binary language of ones and zeros. However, several early electronic computers were not binary computers; they were ternary, with three states, or even quinary, with five. Problematically, the more intermediate states, the harder it became to keep them separate, because disturbances such as power surges, low voltage, or electromagnetic interference would cause mixed signals. The binary was simple to track since it gave distinct signals of "on" and "off." In addition, before the twentieth century, rules and operations to manipulate ones and zeros were already reasoned because an entire branch of mathematics that dealt exclusively with true and false values, Boolean algebra, already existed. Thus, many saw Boolean algebra as the foundation of modern computer science theory. Self-taught English mathematician George Boole developed Boolean algebra in *The Mathematical Analysis of Logic* (1847), in which he presented the truth as systematically and formally represented through logic equations. Unlike the algebra we are familiar with, in Boolean algebra, instead of numbers, the values are true and false; instead of addition, subtraction, multiplication, and division, the Boolean operations are AND, NOT, and OR. For example, if A is true, and B is true, then "A *AND* B" is true.

Conveniently, transistors can be wired in ways to build different types of logic gates and perform these Boolean operations. The assemblage of these logic gates is called the arithmetic logic unit (ALU), which is central to any modern computer. Building computing machines has thus long been viewed as a way to formalize and represent human logical reasoning through material assemblages. Although Boolean algebra laid the foundation for modern computers, the material reality of transistors and electronic circuits eliminates other possibilities regarding how human logic may be systematically and materially formalized. In this regard, quantum computing sheds new light on other aspects of human logic. Quantum computing has gained increased attention over the past 30 years because a qubit (quantum bit) possesses three states; it can be in quantum states, which means a state between 1 and 0 with certain probabilities; but, when measured, a qubit is in the superposition of either 1 or 0 states. Thus, if classical computing uses Boolean logic to erase uncertainty by choosing between 1 and 0, then quantum computing harnesses the power of uncertainty and embraces the ability to be either 1 or 0. For this reason, a quantum computer presents different material assemblages, with the potential to rethink alternative aspects of human logic.

The moral here is that what we perceive as human intelligence is not a sole effort by *human*, but instead is enabled and co-produced by the material world around us. Using human intelligence as a measure for other types of intelligence is thus great hubris, which denigrates the efforts of nonhuman assemblages in the process of co-producing human intelligence. However, if *humans* should no longer be the measure, is there another way to frame intelligence? Recent AI research may provide insights into a non-anthropocentric definition of intelligence.

Generally, there are two types of approaches to building an AI system. The first may be described as an expert system, for which human experts write a rule book for decision-making. For example, if we want to build a plant identification system, we require a group of botanists to list all plant identification rules. This resembles searching for a plant in an encyclopedia. We address multiple questions: is the plant woody or herbaceous? What is the shape of the leaves? What is the shape of the bud? When does it flower? One may create many other questions related to identifying a plant. Many traditional online plant identifier websites are expert systems.

The second approach may be understood as machine learning, for which we need only provide the machine with a large amount of data; the machine itself will reconstruct potential rules for decision-making. Building the same plant identification system with machine learning, we first collect a large number of images of plants labeled with their names. Then, we build an artificial neural network (ANN) and expose this ANN to the plant dataset until it makes predictions with a high success rate. This process is called training. To evaluate how well the model performs, scientists commonly track the prediction error rate using a loss function, which can be optimized through model training. The loss function reflects the AI system's ultimate goal. In a way, all machine-learning problems may be understood as optimization problems; an underlying loss function will always be optimized in order to train a usable model.

Though the concepts of machine learning and AI have existed since the mid-twentieth century, machine-learning techniques did not find currency until the early days of the twenty-first century, due to limited computational power, as well as the amount of data available. With the rise of "big data," as well as increased computing powers, machine learning has produced fruitful results. In 2012, the AI community experienced a major breakthrough in the ImageNet Large Scale Visual Recognition Challenge, a benchmark for computer vision. Computer scientists were challenged to design machine-learning algorithms to perform image recognition; the error rate of the recognition results became a measurement. A team led by Alex Krizhevsky from the University of Toronto designed a model called SuperVision, later known as AlexNet; based on a convolutional neural networks (CNNs) approach, it won the 2012 competition.[6] AlexNet presented a significant drop in the error rate, from 25.8% in 2011 and 28.2% in 2010 to 16.4% in 2012.[7] A mere four years later, in 2016, computer scientists lowered the error rate to only 3%, based on CNNs, while the average error rate for humans doing such tasks was 5%.[8]

Advances in AI research also provide transformative cases for scholars to ask more profound questions about intelligence. Sean Legg and Marcus Hutter, both AI scientists, have proposed the term *universal intelligence* and formalized it through mathematics as a framework with which to compare different forms of intelligence. After reviewing a collection of definitions of intelligence from research groups and organizations, psychologists, and AI researchers,[9] Legg and Hutter ascertained a common thread in these definitions, which involved the interaction of an agent, human or nonhuman, with its environment. Based on this observation and their goal of measuring machine intelligence, they proposed a working definition for universal intelligence: *intelligence measures an agent's ability to achieve goals in a wide range of environments.*[10]

Legg co-founded DeepMind Technologies, later acquired by Google, and was behind the development of AlphaGo and its predecessors; Hutter is a professor and a senior scientist at DeepMind. Their definition of intelligence is deeply rooted in their extensive experience in machine learning, particularly in reinforcement learning (RL), which is the foundation of AI systems like AlphaGo. Essentially, their concept of universal intelligence mirrors an RL framework. In RL, an agent observes the environment's state and takes actions. The effectiveness of these actions is measured by a reward mechanism, typically expressed through a loss function. For instance, in training an RL agent to play Tetris, the agent initially performs random actions such as arbitrarily moving and placing blocks. Certain actions, clearing a line of blocks, for example, result in rewards (i.e., scoring). After thousands of rounds, the agent begins to favor actions that yield higher scores, eventually mastering advanced techniques such as the "Tetris Clear"—clearing four lines simultaneously.[11] Again, any machine-learning technique can be seen as an optimization problem. In RL, the agent strives to optimize its reward function to maximize rewards or minimize punishments. Thus, universal intelligence is a measure of an agent's ability to optimize these reward functions.

Based on this simple agent-environment framework, DeepMind has trained many RL-based AI systems, such as AlphaGo. Only one year after AlphaGo defeated its human competitors, two newer AI systems based on self-play—AlphaGo Master and AlphaGo Zero—overwhelmed the original algorithm. In 2020, MuZero mastered not only one but also four games.[12] The self-play technique implies that the AI system has been playing against itself without any human knowledge input related to the game of Go, except for the basic rules. Not only did these self-play AI systems outperform humans, but they also devised novel strategies that human players had never attempted. To a certain extent, they developed a machine understanding of the game.

In 2019, again using the self-play method, DeepMind developed an AI system called AlphaStar, which attained grandmaster level (the highest rank reachable by competing with other players) in the real-time, strategy video game StarCraft.[13] The AI community regards this experiment as a leap forward because real-time strategy games such as StarCraft are infamous for their infinite combinations of actions based on imperfect information. After watching or playing against AlphaStar, many professional players reported that it had devised new strategies from which the humans could learn; AlphaStar provided new ways to play the game itself. One commentator reported that observing the AI playing the game was akin to watching a drunken kung fu master performing martial arts: awkward, but outrageously effective.[14] Another professional player commented,

> AlphaStar is an intriguing and unorthodox player—one with the reflexes and speed of the best pros but strategies and a style that are entirely its own. The way AlphaStar was trained, with agents competing against each other in a league, has resulted in gameplay that's unimaginably unusual; it really makes you question how much of StarCraft's diverse possibilities pro players have really explored.[15]

These examples could respond to the AI criticism that machines cannot apply knowledge creatively. As we have seen, not only do machines create novel strategies, but importantly, they also develop strategies with a "machine flavor." The implications of universal intelligence extend past training advanced AI systems such as AlphaGo and AlphaStar.

In a way, universal intelligence is a non-anthropocentric definition in which *human* no longer provides the measure; instead, human intelligence becomes an instance of universal intelligence, manifested in the human assemblage. However, it is not yet a posthumanist definition of intelligence because it faces two other challenges: *intelligence as a measurement*, and *individualism.*

Autopoiesis and Artificial Neural Networks

From IQ testing to universal intelligence, measurement has been a primary instrument for questions about intelligence. Thus, the problems associated with "intelligence," such as racism, sexism, classism, and even speciesism, have been products of human exceptionalism in consort with human intelligence as a measurement. For example, critics claim that IQ testing measures nothing but the subject's test-taking skills at the moment of testing; the result, therefore, has little to do with the individual's intelligence. The concept of universal intelligence is confronted with similar critiques, which tie intelligence to metaphysical concepts such as conciseness, soul, and free will, all of which lie beyond measurement.[16] However, Legg and Hutter defend their thesis through a tautology of measurement:

> Our goal is to build powerful and flexible machines and thus these somewhat vague properties are only relevant to our goal to the extent to which they have some measurable effect on performance in some well-defined environment. If no such measurable effect exists, then they are not relevant to our objective.[17]

These critiques miss the point; they realized that the problem lies in measurement, but tying intelligence to metaphysical concepts beyond measurement ironically justifies Legg's and Hutter's approach as a pragmatic choice. Rather than relying on consciousness or soul as *a priori* for intelligence, we need to critique mathematism and formalism embedded in universal intelligence. We must confront what this measured "intelligence" really is. Perhaps formalizing intelligence helps to build powerful machines, but there is no need to rely on this formalized idea to define intelligence. Indeed, if we ponder the video game player's comment above on AlphaStar, the reason the professional players believe this AI to be intelligent is not based on measurement or evaluation. Instead, it is a general impression, a belief that AlphaStar possesses certain potentials that allow it to change and adapt.

What precisely are these abilities that we recognize as intelligence? To answer, we need to detach the concept of intelligence from the urge to test and measure. Only then can universal intelligence become a posthuman concept. Universal intelligence implies agents' abilities to achieve *goals*. However, human scientists define these goals, thus imposing a human standard as to whether a behavior counts as an intelligent move. Ultimately, what universal intelligence measures remains perceived effectiveness from a human perspective, even though the concept itself seems to remove humans from the scale of measurement. It is an intelligence *for* humans rather than intelligence *for* the agent itself.

I will employ the concept of autopoiesis to elucidate the meaning of "goal-directedness" in the context of measuring intelligence. In the influential paper "What the Frog's Eye Tells the Frog's Brain," Chilean biologist and philosopher Humberto Maturana, along with his colleagues, revealed that a frog's eye does not simply capture and transmit an exact image to its brain for analysis. Instead, the process involves four distinct groups of nerve fibers, each responsible for a specific type of operation on the visual data. These fibers first process the image and then relay this organized information to the brain. This finding contests the prevalent belief that the nervous system functions primarily as a tool for gathering information to construct an internal representation of the external environment. On the contrary, the environment triggers a set of operations in the nervous system that produce reality inside the organism. Autopoiesis turns the environment outside-in. Based on this biological observation, Maturana and Francisco Varela, also a Chilean biologist and philosopher, argue that:

> [w]e as observers have access both to the nervous system and to the structure of the environment. We can thus describe the behavior of an organism as though it arose from the operation of its nervous system with representations of the environment or as an expression of some goal-oriented process. These descriptions, however, do not reflect the operation of the nervous system itself. *They are good only for the purpose of communication among ourselves as observers.* They are inadequate for a scientific explanation.[18]

The environment merely *triggers* the system to operate and construct a reality of its own. When folding the environment and reality outside-in, concepts such as consciousness, soul, and free will become epiphenomena produced through human system operations. This could be understood as a version of radical constructivism. Goal-directedness becomes an epiphenomenon, too, for humans to describe causal relationships. To use the logic of goal-directedness to describe how and why other systems operate is the privilege of the human observer. We explain system operations in a manner that fits within the descriptive category of goal-directedness, which now becomes a tactic of observation. We choose to believe that goals exist, then observe phenomena that fit within the framework of goal-oriented behaviorism.

However, as Maturana and Varela note, evaluating and measuring goal-directed phenomena has nothing to do with system operations themselves. I will analyze past examples of AI biases with autopoiesis theory to further illustrate what this means. One notable case involved an image-recognition model that erroneously identified husky dogs as wolves. This error was traced back to the training dataset, where images of wolves often possessed snowy backgrounds. The model, therefore, learned to associate snow with wolves, leading to confusion when presented with images of huskies against snowy backgrounds.[19] Other systems are gender-biased. In one example, an image-recognition AI system tended to label human figures at the forefront of a kitchen background as female.[20] In Midjourney, a generative AI system that produces images, if a user gave the prompt "a person in a garden," Midjourney would generate a white woman or a feminine figure in front of a garden scene (Figure 8.1).

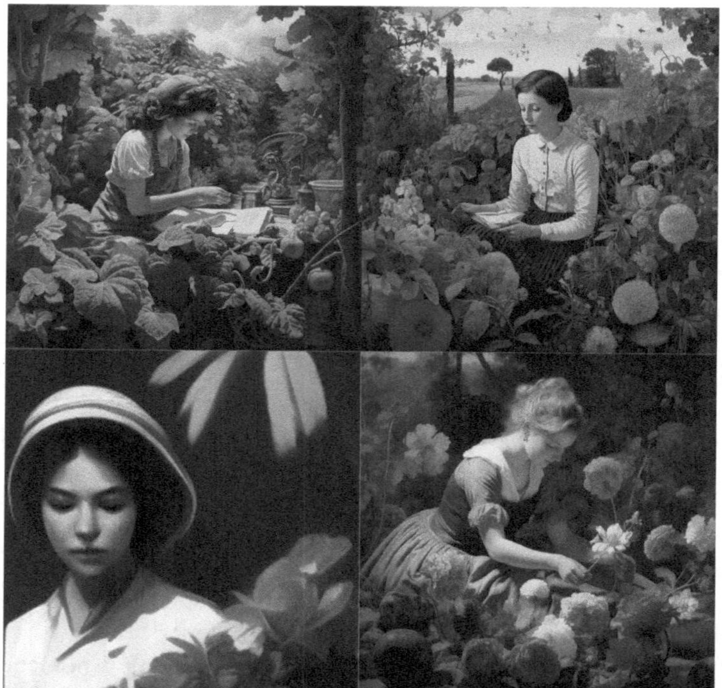

FIGURE 8.1 Gender-biased images generated with Midjourney. Images generated by the author through Midjourney.

In these examples, there existed inherent biases in the training data because the images used for training were merely culled from the internet, and thus reflected society's persistent gender stereotypes. If one searches "people gardening," Google's image search will return many images featuring white women gardening. Moreover, these images are likely to be used to train the above AI systems. Further examples abound, and have been used to challenge scientists to create fair and "unbiased" AI.[21] Yet, in early 2024, Google's Gemini created some historically inaccurate images, such as Nazi-era German soldiers as people of color, possibly due to an overcorrection of AI racial bias.[22] Training "unbiased AI" is apparently more complex than we think.

However, the analytical values of these cases were not discussed fruitfully by computer scientists and critics, especially within an autopoiesis framework. If these AI systems are understood as autopoietic systems, then their environments are constructions of their own system operations, which are entirely different from human constructions. If a human is presented with an image of a figure in a kitchen setting and asked to identify the figure's gender, they would ignore the background and turn their attention to the figure itself. However, the machine-learning architecture for most image-recognition algorithms is called convolutional neural networks (CNN). In ANN, artificial neurons are constructed in layers. In CNN, particularly, a convolutional layer acts as a "filter" that passes over the image, scanning several pixels at a time from top to bottom, and left to right, and creating a feature map. Then, a pooling layer checks this feature map and abstracts it into small edges that represent the object in the image. Finally, fully connected layers would make a prediction based on these small edges. The training process uses training data to fine-tune each neuron. A CNN is therefore a perfect example of an autopoietic system; after training, the CNN is hard-wired such that the environment only triggers a set of neural activations. A saying in the AI community notes, "Neurons that fire together, wire together." A CNN does not absorb an image as a whole but as pieces; it is as if a person "reading" a picture used a magnifying glass to search for the small edges that define objects. Using this magnifying glass analogy, a person born holding a magnifying glass would be exposed to a completely different "environment" than the rest of us. Just as Matuana's frog constructs a reality with four groups of fibers, a CNN constructs a reality through a combination of ANN layers.

Based on CNN architecture, scientists combine multiple and various layers for building hierarchies in feature complexity, thus constructing deep convolutional neural networks (DCNN). In DCNN, the first few layers detect edges, the middle layers detect portions of the object, and the final layers detect the object itself. Neuroscientists have found that the increasing feature complexity of DCNN resembles the increasing complexity that occurs in visual object recognition in humans.[23] Others found that the activation of DCNN was similar to that of gamma-band activity within the human visual cortex.[24] Many have duly argued that DCNN resembles the method by which humans see, and some regard DCNN as a biologically inspired approach. These findings seem to contradict autopoiesis theory by drawing similarities between machines and human system operations. On the contrary, these examples prove that DCNN does not resemble humans at all because they perform an operation that merely resembles the *human cortex*, which is only a part of the cognitive assemblage we call human.

We may explain the behavior of a CNN as a goal-directed behavior, but this explanation has nothing to do with system operations; AI systems will execute what we *say* rather than what we *mean*. After training, CNN will label an image, but this action does not equal image recognition. We have no idea what CNN maintains as goals for itself or if *goal* is even a concept within CNN's reality as constructed by its system operation.

The AI community has struggled with the disconnect between goal and system operation. Autopoiesis sheds new light on problems in AI safety research, such as the problem of reward hacking. In an amusing example, a simulated AI robotic arm was trained to flip a pancake; scientists set up a reward function such that the robot would receive a small reward for each second that the pancake did not hit the floor. This sounded logical since a human would continuously flip the pancake to keep it in the air as long as possible. However, the robot arm hurled the pancake into the air with as much force as possible, to maximize the reward.[25] Articulating reward goals and AI behaviors is a constant task in AI research because AI systems may exploit the reward function by performing unexpected behaviors.[26] Powerful AI systems, such as AlphaGo or MuZero, might be understood as gratifying accidents, where the articulated reward goal matches the system operation. This synergy between goals and system behaviors points us in another direction: we can frame intelligence with ideas of co-production and co-evolution.

To consider synergy between scientists and AI systems, we must practice what Niklas Luhman calls *second-order observation*, or the practice of observing others observe. New insights arise when we bring the observer into the equation. When intelligence is tied to measurement, it becomes a product of observation; thus, an observer plays an essential role in conceptualizing intelligence—how intelligent an entity appears to the observer at the moment of observation. In human–robot interactions, *perceived intelligence* is a key concept. Robot engineers know the underlying rules that govern how a robot would behave in a given circumstance; to them, the robot's behaviors are transparent and predictable, thus non-intelligent. However, for a non-expert without prior knowledge of the robot's underlying rules, its behaviors may appear intelligent. Perceived intelligence was used to mediate the knowledge gap between engineers and non-expert users. To make robot behaviors appear more intelligent, engineers would intentionally program randomness so that the behavior patterns were less predictable. However, after many interactions, users will eventually detect patterns in these limited random behaviors, and decide that the robot is not intelligent, after all. Thus, a popular approach in human–robot interaction is the so-called Wizard of Oz trick, in which a human hides behind the robot and conducts a conversation with another human to create the illusion of intelligence.[27]

What separates these two situations—the Wizard of Oz versus limited random actions? We may conclude that users find the robot unintelligent because its behaviors become predictable over time. However, we could draw a different conclusion if we absorb observers into the equation: the users become more intelligent over time, discerning the robot's deceptive behaviors. The robot is unintelligent because it cannot co-evolve with the users, whereas another human hiding behind the robot adapts to the users and advances the conversation.

Similarly, if we reconsider the Turing test, the seemingly innocent discriminator becomes a critical player in the imitation game because the game eventually becomes a test for all parties. Recursive observation and learning between observer and participants will force them to adapt to each other and become more intelligent in playing the imitation game. Measurement overlooks the intelligence emerging from dynamic feedback loops between the observer and the observed. From this perspective, intelligence points to a direction on the line of co-production and co-evolution within an assemblage framework. Or, to use Donna Haraway's term, *sympoiesis*—making with.[28]

Sympoiesis and the Emergence of Intelligence

With current AI advancement, we lack vocabularies and concepts for a non-individualistic view of intelligence, in both theory and practice, because AI was originally conceived as an autonomous machine. Collaborative intelligence is a concept proposed in AI research to challenge the rivalry between humanity and AI constructed over decades by popular culture and media. It posits that a synergy between people and AI is the basis for considering machine intelligence.[29] We should extend this argument to include all types of human and nonhuman agents, and construct a posthumanist definition of intelligence as the basis for AI research.

Computer scientists regard an agent as a black box with input and output; similarly, researchers regard intelligence as a person's capacity in IQ testing. Assemblage thinking exists at odds with the inherent individualism of how intelligence is conceived—a quality of a pre-existing agent. However, as I have explored in previous chapters, an agent is simply a product of observation to attribute perceived effectiveness distributed *within* and *around* the so-called agent. It is merely a carrier of intelligence that has emerged from dynamic co-evolution and co-production processes involving actors more than the agent itself.

One way to observe the emergence of intelligence is to construct an arbitrary diagram of an agent and its outside environment, and then analyze both inside and outside the agent. First, intelligence is an emergent property of the interactions between the components that give rise to the assemblage. For example, swarm intelligence points to the collective behavior of decentralized, self-organized systems, such as a colony of ants. While exploring their environment, ants leave pheromones to guide each other to resources. The behavior of an ant appears simple, but if each ant repeats the same actions, a colony of ants appears remarkably efficient in exploring its environment. This behavior inspired a famous instance in computer science, called the ant colony optimization algorithm, which has proven useful in pathfinding optimization problems.[30] This type of intelligence, also known as "hive mind," occurs where intelligence is not located within any of the individuals, but as a property that emerges from the interactions of each part.

In a way, all forms of intelligence—human, machine, or animal—should be understood as emergent phenomena. Human and animal intelligence are nothing but electrochemical interactions between neurons and a distributed nervous system. Machines are the same; as we saw in the above CNN example, in ANNs, AI is achieved through the interactions of layers of interconnected artificial neurons, or functions between input and output. An AI system is thus a conglomeration of connected functions.

Second, outside the boundary of the agent, the term *mediation* may help us conceptualize that an agent's intelligent behaviors are constantly mediated by other assemblages. The mediation may be understood through Andy Clark's "extended mind" theory, which reminds us that the effectiveness of human intelligence is achieved by off-loading human cognitive functions onto other objects in the environment. Human intelligence becomes a continually changing phenomenon, depending on the temporary and ephemeral assemblage at the moment of observation. A paper-based IQ test evaluates the effectiveness of finishing the questions by the temporary assemblage of human, pen, paper, and perhaps corrective lenses; all factors contribute to how well the subject performs at the test-taking moment. We must also account for the breakfast eaten that morning, as well as the diverse bacteria and enzymes in the test-taker's digestive system, which help process food into energy for the electrochemical nerve impulses that give rise to "intelligent behaviors."

On the other hand, we may add a time scale to the mediation process and see its effects as co-evolution and co-production. In the story of transistors and Boolean algebra, the users of the humanoid robots and the discriminator in the Turing test are all instances where the human and the machine become each other's media and co-evolve over time. We need to reframe intelligence as a collection of observed phenomena that is, in fact, a result of random interactions in co-production and co-evolution among assemblages.

To understand intelligence based on the inside and outside of the agent is arbitrary. In fact, it is crucial to recognize that reality is a continuous mesh of heterogeneous bodies within the assemblage framework. Random interactions and feedback loops exist between different agents, and if certain interactions become useful for a temporary assemblage, then synergies and loose couplings would happen among these bodies. Such thinking is found in many contemporary feminist materialists' writing. As Jane Bennett noted, "bodies enhance their power in or as a heterogeneous assemblage."[31] This process is also essential to Donna Haraway's notion of *sympoiesis* (making with); nothing is truly autopoietic and self-organizing, and things are always "making with" others.[32] Sympoiesis produces intelligence.

In a posthumanist, sympoietic framework, intelligence becomes a shared ability among different assemblages to exchange effects through recursive observation and learning, thus forming synergies and couplings to gain power. Intelligence speaks to the process of attunement and symbiosis between assemblages, as in the moment when computer scientists' goals accidentally match operations in an AI system. We can frame this phenomenon as "intelligence of co-production" or "co-productive intelligence." What we perceive as intelligence—whether human, machine, or animal—becomes a specific instance of observed local manifestation of co-productive intelligence distributed across the heterogeneous landscape of human and nonhuman, biotic and abiotic assemblages.

Co-Productive Intelligence

To summarize this part, I propose three types of co-productive intelligence as an attempt to depart from how intelligence is currently discussed: (1) intelligence in adversarial relations, (2) intelligence in symbiosis (symmetrical and asymmetrical), and (3) intelligence in loose coupling.

Adversarial intelligence gestures to competition in long-term evolution, in which complex behaviors and responsive strategies emerge. In examples like AlphaGo and AlphaStar, humans and AI agents participated in competitive environments, and new strategies emerged; together, humans and machines expanded the possibilities of a game.

In recent years, the AI community has exploited adversarial relations and achieved promising results. Many machine-learning approaches involve two adversarial parts training together. One example is that of generative adversarial networks (GANs). Before diffusion models (e.g., DALL-E.2 and Midjourney) took over image generation in mid-2022, GANs were the go-to solution for computer scientists. There, GANs might be demonstrated with a forgery example. Imagine two neural networks, one generator (forger) and one discriminator. The forger's goal is to produce photorealistic images to deceive the discriminator, and the goal of the discriminator is to ascertain whether it is a real image. The two networks are trained simultaneously until the discriminator cannot distinguish a fake image from a real image. Researchers have used this technique to train AI systems to render photorealistic images and stylized paintings.[33] In 2018, GANs was named one of the "10 breakthrough technologies" by the Massachusetts Institute of Technology (MIT); a commenter noted that GAN "gives machines something akin to a sense of imagination, which may help them become less reliant on humans."[34]

Even though diffusion-based systems, including DALL-E.2 and Midjourney, quickly superseded GAN-based image generators, the concepts embedded in GANs are not completely obsolete. For example, Variational Autoencoder (VAE), a key component of the neural network model used in Stable Diffusion to improve the quality of AI-generated images, essentially assumed two neural networks (encoder and decoder) in an adversarial relationship. Similarly, in RL, Deep Deterministic Policy Gradient (DDPG) is a technique that establishes an actor-critic architecture within the learning agent, in which the actor decides which action to take in an environment, and the critic evaluates the action and informs the actor how effective the action was and how it could be improved.

Moreover, computer scientists have found that competitive multi-agent techniques in RL not only train models faster but they also give rise to behaviors far more complex than those allowed by the environment alone.[35] In an example developed by OpenAI, two agents were asked to play hide-and-seek in a simulated environment with walls, boxes, and boards. After training, both agents (seeker and hider) learned to use these tools to their advantage and began to develop methods to exploit the environment in which they were trained. One agent learned to stand on a box and apply force to the agent itself; it then "surfed" the boxes and jumped over walls.[36]

Intelligence in symbiosis or structural coupling is the second type of coproductive intelligence. The evolution from prokaryotic cells to eukaryotic cells has been widely discussed among posthumanism scholars.[37] It is now believed that mitochondria were once a type of bacteria before being coopted as permanent organelles of cells in which they were originally parasites. It seems that adversarial relationships may transform into symbiotic relationships where two or more competing assemblages self-synchronize in terms of their inputs and outputs so that they can exchange effects. We may find the relationship of symbiosis in many plant communities. For example, root nodules are primarily found on the roots of legumes, or the pea family (*Fabaceae*), including peas and soybeans, as well as on trees such as the black locust (*Robinia pseudoacacia*). These root nodules are the result of legumes forming a symbiosis with nitrogen-fixing bacteria that helps convert dinitrogen (N_2) from the atmosphere into ammonia (NH_3), which can then be used by plants. In ANNs, neurons (non-linear functions) are structurally coupled in layers and networks. One neuron (function) cannot do much, but structurally coupled neural networks become intelligent.

Furthermore, OOO and other posthumanist ideas expand symbiosis beyond biology as a metaphor for how living entities form symbiotic relations with other assemblages, including non-material organizations, institutions, and historical objects.[38] Symbiotic relationships involve *structural coupling*, first discussed by Maturana and Varela when introducing autopoiesis theory. Levi Bryant provides a more accessible explanation of structural coupling by describing it as "a relation in which one or two entities are dependent for stimuli or flows from one another in order to engage in their own operations and becoming."[39] Bryant divides structural coupling into bidirectional and unidirectional; this is similar to Harman's symmetrical and asymmetrical symbiosis. Symmetrical symbiosis, or bidirectional coupling, occurs when both assemblages or systems require flows from the other to engage in a joint operation, such as mitochondria acting as permanent organelles in living cells. In asymmetrical symbiosis, or unidirectional coupling, one system depends entirely on the other; an ecosystem type may be said to be unidirectionally coupled with the climate in its region. Asymmetrical symbiosis or unidirectional coupling as a metaphor speaks to adaptive strategies which a system develops to take advantage of another system. In symbiosis or structural coupling, intelligence manifests as the ability to link one system to another so that its own system operation becomes assimilated into another system's operation; both systems benefit from the assimilation of system operations, and both gain power.

Intelligence in loose coupling is the third type of co-productive intelligence. Computer scientists and system engineers use coupling to describe the interdependence between different programming objects or systems. In programming, there are two types of coupling—tight and loose—with the latter preferred for greater flexibility so that altering one programming object or system is unlikely to affect another. Similarly, sociologist Niklas Luhmann, who used structural coupling to describe social systems, argued that "[l]oosely coupled systems are more stable than tightly coupled ones. 'Tight coupling' is a very improbable arrangement. It is not to be found in nature."[40] Many scholars disagree with Luhmann's claim regarding loosely coupled systems.[41] Symmetrical symbiosis may be understood as a type of tight coupling, and symmetrical symbiosis is found nearly everywhere in "nature." Certainly, in symmetrical symbiosis, if one of the tightly coupled systems (for example, the soybean) is destroyed, then the other tightly coupled system (here, nitrogen-fixing bacteria) can no longer survive. Viewing loose coupling as beneficial is based on its resistance to external disturbance, overlooking the intensive exchange of flows and functions in tightly coupled systems that do not exist in loosely coupled systems. Instead of placing a high value on loose coupling, it can be understood as another type of relationship from which intelligence emerges.

Loose coupling is a type of relationship we commonly encounter, such as the relationship between computer scientists and machine-learning algorithms. The synergies between these assemblages could be seen as "happy accidents" when scientists' goals match the observed system operation in an AI algorithm, and thus effectiveness comes into being.

These three types of co-productive intelligence attempt to create a set of vocabularies with which one could consider the role of intelligent machines within a posthumanist, sympoietic framework. I will illustrate this point further with Sougwen Chung's artwork "Drawing Operations," a series of performances in which the artist drew alongside robots (Figure 8.2).[42] In the first version, the artist drew alongside a robot arm called D.O.U.G_1 (Drawing Operations Unit: Generation_1). D.O.U.G_1 mimicked the artist's movements by analyzing her drawing gestures through an overhead camera and reproducing them. The final artifact is a co-produced drawing through collaboration. However, the robot's movements were not perfect reproductions of the artist's gestures. Although the algorithm tracked the artist's linework in the digital simulation, the movements were dramatically altered when translated to a robotic arm.

Imperfection was unavoidable in this loosely coupled system. The computationally heavy, live, computer-vision analysis created an inevitable system

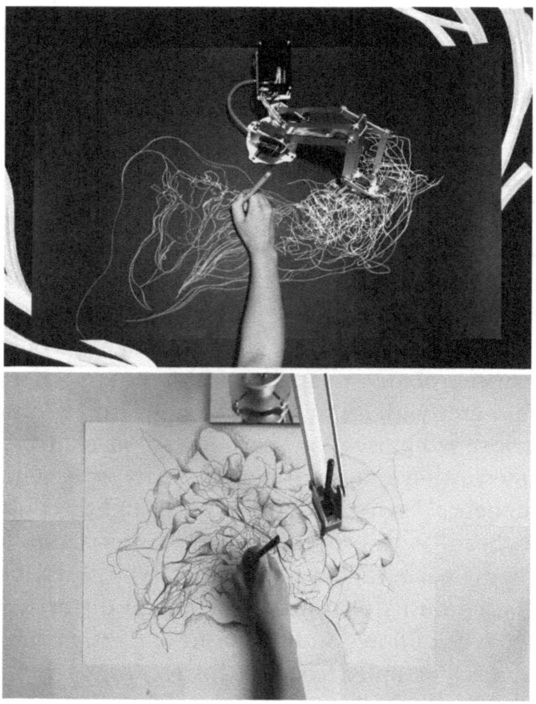

FIGURE 8.2 Top: "Drawing Operations" (2015); bottom: "Drawing Operations" (2017). Courtesy of Sougwen Chung.

latency, creating a lag between the artist's movements and the robot's reproduction of them. The linework quivered as if the robotic hand was unsteady. In real-time, the artist was forced to adapt her movement to the robot's. Her new gestures fed back to the robot, and the robot produced a new set of movements to which the artist was again compelled to adapt. The artist and the robot formed a positive feedback loop; together, they became a loosely coupled system with new styles and techniques manifested in the co-produced artifact, gradually synchronizing and attuning to each other in a territory alien to both.

Later, the D.O.U.G_2 was designed around the notion of memory. Chung and her team deployed machine-learning techniques to teach the artist's drawing style to an AI system. They fed an algorithm with decades of the artist's work so that the AI would attempt to reproduce Chung's drawing patterns. The machine subsequently developed its own understanding of the artist's style, and expressed a machine interpretation in their later drawing collaborations.

Chung's art installations raise questions about perceived errors and glitches in systems, reframing the notion of failure as an important factor in art production and design. As the artist noted, "The robot mimics the artist like a partner in an improvisational performance. It is an AI that embraces every glitch, bug, and error."[43] From a posthumanist perspective, these so-called glitches, bugs, and errors are fundamental facets of operation defined by an algorithm and the physical armature of a robotic arm. To human eyes, they may appear to be errors, but they are the ways in which the robotic systems operate within the environment. In theory, we might minimize system latency by using a faster processing unit and by accounting for the robotic arm's physical limitations. However, if one repairs all the glitches, bugs, and errors so that the robot perfectly replicates every detail of the artist's gestures, the art does not exist. The results become anticipated and predictive, while Chung's art lies in the unexpected outcomes of human–robot collaboration. When D.O.U.G_2 learned the artist's style through her works, there were no right or wrong answers for interpretation because they were merely machine interpretations. If the machine replicates the artist's drawing style perfectly through learning, then the machine hardly differs from a photocopier. Thus, adaptation between the artist and the robot entails exploiting each other's limitations and errors. Or, to use Chung's words, her artwork is about "embracing the imperfections and recognizing the fallibility of both human and machine in order to expand the potential of both."[44] As discussed earlier, the foundation for the emergence of intelligence is co-production between assemblages. In Chung's work, intelligence manifests in the loose coupling and attunement between humans and machines. In this attuning process, the potentials and possibilities of the human-machine assemblage expand.

In the cybernetic environment, our relationship with machines must not be limited to the co-dependent relationship between users and tools. Instead, artistic explorations suggest that potential may be found in interdependent relationships, where new questions arise, new understandings form, and new strategies emerge. With the conviction that machines are not tools for automation but actors that have been involved in the co-production of the environment, we may build an alternative way to understand failures and errors when collaborating with intelligent machines. When machines produce unexpected strategies, those are not necessarily errors. They are, instead, opportunities for us to form different understandings of the questions posed, just as AlphaGo and AlphaStar have expanded players' understanding of their games.

A Posthumanist Perspective on AI Safety and Bias

I will end this chapter by providing notes on AI safety and bias from a posthumanist perspective because I have not read such analyses in my research. Viewing machine intelligence as a threat to humanity is a popular sentiment, especially after the twentieth century's two world wars, when society witnessed the potential destructive power of technology. Popular culture also aided the growth of this sentiment, since AI often assumes a supervillain role in dystopian novels and movies. Imagining rogue AI is a form of the technological sublime, in which we entertain a potential threat from a safe position. We comprehend the threat with human reasoning and rationality, which, according to Kant, is the basic structure of sublimity. However, this view is deeply rooted in an anthropocentric understanding of human intelligence and agency. Recognizing the sublime quality of machines extends the anthropocentric understanding of human agency. In turn, it adds to a sense of human hubris, since entertaining the potential threat of machines ironically reflects a sense of human control and mastery over machines.

What we understand as human agency has always been distributed in a heterogeneous landscape of human and nonhuman components. These components are biological and technological, including tools, machines, technological systems, models, algorithms, and ANNs. We have always been cyborgs, and we have always explored the environment in company with other actors. Machines have helped us expand our understanding of the environment, and the rejection of their perspectives of the environment is rebuffing what it means to be human. Thus, embracing machine perspectives is to accept the cyborg condition of being human.

I argue that adopting a posthumanist understanding of intelligence could greatly expand the discourse of research on AI safety and bias. Over the years, algorithms trained by AI research clusters, such as Google and Meta, have resulted in "biased" AI systems that uncannily resemble humanity's biases. In 2015, a Google image-recognition algorithm placed the label "gorilla" on Black people. In 2016, searching female names, such as "Stephanie Williams," on LinkedIn triggered search suggestions asking if the user sought similar male names, such as "Stephen Williams." That same year, Tay, a Microsoft chatbot, spent a day learning from Twitter, after which it began to tweet racist and sexually-charged messages. Other examples include gender-biased image-recognition and -generation algorithms, as we have seen earlier in this chapter. As we now recognize, AI biases are reflected from everyday gender and racial stereotypes.

According to their own accounts, Google and Facebook quickly "fixed" these "biased" algorithms after the public launched scathing criticisms. Researchers wanted to create "unbiased" AI algorithms. However, a deeper irony lurks behind these stories. From a posthumanist perspective, these AI algorithms are not biased at all, because they hold true to the datasets on which they were trained by cruelly reflecting an ugly aspect of humanity which many of us dare not recognize. In a way, to teach an algorithm to be "racial-neutral" equates to teaching someone to see no skin color. Yet "seeing no color" is a problem in itself, one of society's ignorance of the reality of systemic racism.

At the center of this irony is that we seek to treat AI algorithms as automation machines, instead of as yet more voices on issues we thought to address solely as humans. If we focus on the bigger picture, these blunt and "biased" AI systems played an important role in recognizing our biases in online spaces such as Twitter and LinkedIn. There is a saying in the machine-learning community: "Garbage in, garbage out." Yet, a garbage AI reveals how the internet has become a hoarding place for humanity's mental garbage.

A similar argument can be made concerning image generators that have disrupted creative professions. The work "Théâtre D'opéra Spatial" dropped a bomb onto the art professions that brokered a war between artists and AI image generators. Created using Midjourney, Jason Allen submitted this work to the Colorado State Fair's annual art competition using the name "Jason M. Allen via Midjourney." He won the blue ribbon for emerging digital artists. While Allen unapologetically boasted of his achievement, the news set off a backlash that eventually led to a mass protest of artists against AI-generated artwork. Because AI researchers scraped the internet for training data, including images by these protesting artists, the trained AI system would generate similar visual styles. Although rules and regulations concerning training data in AI research must be encouraged, a posthumanist twist is needed to bypass the inherent human-machine rivalry narratives. To some degree, the boycotting human artists must have felt insecure when many non-artists exclaimed that AI image generators would one day replace them. Such a comparison is based on an anthropocentric view, in which human artists become the measure of creativity and imagination. From a posthumanist perspective, the goal of building powerful machines should never be about replacing or replicating human capacity, but about providing alternative, non-anthropocentric perspectives to expand our all-too-human pathways of thinking and making.

Notes

1 Merriam-Webster, "Definition of INTELLIGENCE." https://www.merriam-webster.com/dictionary/intelligence
2 Turing, "Computing Machinery and Intelligence."
3 Premack, "Human and Animal Cognition."
4 Trewavas, "Aspects of Plant Intelligence."
5 Belkhir, "Race, Sex, Class & 'Intelligence' Scientific Racism, Sexism & Classism"; Dennis, "Social Darwinism, Scientific Racism, and the Metaphysics of Race"; McNally, "Scientific Racism and The Politics of Looking."
6 Krizhevsky, Sutskever, and Hinton, "Imagenet Classification with Deep Convolutional Neural Networks."
7 Russakovsky et al., "ImageNet Large Scale Visual Recognition Challenge."
8 Langlotz et al., "A Roadmap for Foundational Research on Artificial Intelligence in Medical Imaging."
9 Legg and Hutter, "A Collection of Definitions of Intelligence."
10 Legg and Hutter, "Universal Intelligence."
11 See https://www.youtube.com/watch?v=iI6TQOQ_Ccc for an example of a RL-based agent playing Tetris. Last accessed: April 29, 2024.
12 Schrittwieser et al., "Mastering Atari, Go, Chess and Shogi by Planning with a Learned Model."
13 Vinyals et al., "Grandmaster Level in StarCraft II Using Multi-Agent Reinforcement Learning."
14 Two Minute Papers, "DeepMind's AlphaStar: A Grandmaster Level StarCraft 2 AI - YouTube." https://www.youtube.com/watch?v=jtlrWblOyP4. Last accessed: April 29, 2024.
15 The AlphaStar Team, "AlphaStar." https://deepmind.com/blog/article/alphastar-mastering-real-time-strategy-game-starcraft-ii. Last accessed: April 29, 2024.
16 Legg and Hutter, "Universal Intelligence."
17 Legg and Hutter, 42.
18 Maturana and Varela, *The Tree of Knowledge*, 129. Emphasize added.
19 Ribeiro, Singh, and Guestrin, ""Why Should I Trust You?"
20 Zhao et al., "Men Also like Shopping."
21 Computer scientists tend to avoid difficult social science questions regarding gender equality and justice; most of the time, unbiased AI means reflecting whatever the training data represents, excluding issues of gender equality outside the scope of the research. However, this "unbiased" attitude limits the potential for an AI system to be used creatively, for activism.
22 See a media report here: https://www.theverge.com/2024/2/21/24079371/google-ai-gemini-generative-inaccurate-historical. Last accessed: April 29, 2024.
23 Kriegeskorte, "Deep Neural Networks."
24 Kuzovkin et al., "Activations of Deep Convolutional Neural Networks Are Aligned with Gamma Band Activity of Human Visual Cortex."
25 See a video here:https://dzamqefpotdvf.cloudfront.net/p/images/2cb2425b-a4de-4aae-9766-c95a96b1f25c_PancakeToss.gif._gif_.mp4. Last accessed: April 26, 2024.
26 A list of unexpected AI behaviors https://docs.google.com/spreadsheets/d/e/2PACX-1vRPiprOaC3HsCf5Tuum8bRfzYUiKLRqJmbOoC-32JorNdfyT-

iRRsR7Ea5eWtvsWzuxo8bjOxCG84dAg/pubhtml. Last accessed: April 26, 2024.

27 Bartneck et al., "Measurement Instruments for the Anthropomorphism, Animacy, Likeability, Perceived Intelligence, and Perceived Safety of Robots."
28 Haraway, *Staying with the Trouble.*
29 Epstein, "Wanted: Collaborative Intelligence."
30 Dorigo and Di Caro, "Ant Colony Optimization."
31 Bennett, *Vibrant Matter: A Political Ecology of Things*, 23.
32 Haraway, *Staying with the Trouble.*
33 Goodfellow et al., "Generative Adversarial Networks."
34 MIT Technology Review, "2018 MIT Technology Review." https://www.technologyreview.com/10-breakthrough-technologies/2018/. Last accessed: April 29, 2024.
35 Bansal et al., "Emergent Complexity via Multi-Agent Competition."
36 https://www.youtube.com/watch?v=kopoLzvh5jY&ab_channel=OpenAI. Last accessed: April 29, 2024.
37 Authors such as Manual DeLanda, Graham Harman, and Levi Bryant all use this example to demonstrate the process of forming assemblages.
38 Harman, *Object-Oriented Ontology*, 112.
39 Bryant, *Onto-Cartography: An Ontology of Machines and Media*, 153.
40 Luhmann, *Introduction to Systems Theory*, 123.
41 Probert, "Book Review: Introduction to Systems Theory."
42 Drawing Operations, 2015; Drawing Operations, 2017
43 Chung, *Drawing Operations*, 2015.
44 Watch her full TED Talk here: https://www.youtube.com/watch?v=q-GXV4F-d1oA. Last accessed: April 29, 2024.

Bibliography

Bansal, Trapit, Jakub Pachocki, Szymon Sidor, Ilya Sutskever, and Igor Mordatch. "Emergent Complexity via Multi-Agent Competition." *arXiv:1710.03748 [Cs]*, March 14, 2018. http://arxiv.org/abs/1710.03748.

Bartneck, Christoph, Dana Kulić, Elizabeth Croft, and Susana Zoghbi. "Measurement Instruments for the Anthropomorphism, Animacy, Likeability, Perceived Intelligence, and Perceived Safety of Robots." *International Journal of Social Robotics* 1, no. 1 (January 2009): 71–81. https://doi.org/10.1007/s12369-008-0001-3.

Belkhir, Jean. "Race, Sex, Class & 'Intelligence' Scientific Racism, Sexism & Classism." *Race, Sex & Class* 1, no. 2 (1994): 53–83.

Bennett, Jane. *Vibrant Matter: A Political Ecology of Things*. Durham, NC: Duke University Press, 2010.

Bryant, Levi R. *Onto-Cartography: An Ontology of Machines and Media*. Speculative Realism. Edinburgh: Edinburgh University Press, 2014.

Chung, Sougwen. *Drawing Operations*. 2015. Mixed media. https://sougwen.com/project/drawing-operations.

———. *Drawing Operations*. 2017. Mixed media. https://sougwen.com/project/drawingoperations-memory.

Dennis, Rutledge M. "Social Darwinism, Scientific Racism, and the Metaphysics of Race." *The Journal of Negro Education* 64, no. 3 (1995): 243–52. https://doi.org/10.2307/2967206.

Dorigo, M., and G. Di Caro. "Ant Colony Optimization: A New Meta-Heuristic." In *Proceedings of the 1999 Congress on Evolutionary Computation-CEC99 (Cat. No. 99TH8406)*, 2 (1999): 1470–77. https://doi.org/10.1109/CEC.1999.782657.

Epstein, Susan L. "Wanted: Collaborative Intelligence." *Artificial Intelligence* 221 (April 2015): 36–45. https://doi.org/10.1016/j.artint.2014.12.006.
Goodfellow, Ian J., Jean Pouget-Abadie, Mehdi Mirza, Bing Xu, David Warde-Farley, Sherjil Ozair, Aaron Courville, and Yoshua Bengio. "Generative Adversarial Networks." *arXiv:1406.2661 [Cs, Stat]*, June 10, 2014. http://arxiv.org/abs/1406.2661.
Haraway, Donna J. *Staying with the Trouble: Making Kin in the Chthulucene*. Durham, NC: Duke University Press, 2016.
Harman, Graham. *Object-Oriented Ontology: A New Theory of Everything*. London: Penguin, 2018.
Kriegeskorte, Nikolaus. "Deep Neural Networks: A New Framework for Modeling Biological Vision and Brain Information Processing." *Annual Review of Vision Science* 1, no. 1 (2015): 417–46. https://doi.org/10.1146/annurev-vision-082114-035447.
Krizhevsky, Alex, Ilya Sutskever, and Geoffrey E. Hinton. "Imagenet Classification with Deep Convolutional Neural Networks." In *Advances in Neural Information Processing Systems*, 1097–1105, 2012.
Kuzovkin, Ilya, Raul Vicente, Mathilde Petton, Jean-Philippe Lachaux, Monica Baciu, Philippe Kahane, Sylvain Rheims, Juan R. Vidal, and Jaan Aru. "Activations of Deep Convolutional Neural Networks Are Aligned with Gamma Band Activity of Human Visual Cortex." *Communications Biology* 1, no. 1 (August 8, 2018): 1–12. https://doi.org/10.1038/s42003-018-0110-y.
Langlotz, Curtis P., Bibb Allen, Bradley J. Erickson, Jayashree Kalpathy-Cramer, Keith Bigelow, Tessa S. Cook, Adam E. Flanders, et al. "A Roadmap for Foundational Research on Artificial Intelligence in Medical Imaging: From the 2018 NIH/RSNA/ACR/The Academy Workshop." *Radiology* 291, no. 3 (June 2019): 781–91. https://doi.org/10.1148/radiol.2019190613.
Legg, Shane, and Marcus Hutter. "A Collection of Definitions of Intelligence." *arXiv:0706.3639 [Cs]*, June 25, 2007. http://arxiv.org/abs/0706.3639.
———. "Universal Intelligence: A Definition of Machine Intelligence." *Minds and Machines* 17, no. 4 (December 1, 2007): 391–444. https://doi.org/10.1007/s11023-007-9079-x.
Luhmann, Niklas. *Introduction to Systems Theory*. 1 edition. Cambridge; Malden, MA: Polity, 2012.
Maturana, Humberto R., and Francisco J. Varela. *The Tree of Knowledge: The Biological Roots of Human Understanding*. The Tree of Knowledge: The Biological Roots of Human Understanding. Boston, MA,: New Science Library/Shambhala Publications, 1987.
McNally, Cayla. "Scientific Racism and The Politics of Looking." In *Jordan Peele's Get out: Political Horror*, edited by Dawn Keetley, 212–22. New Suns: Race, Gender, and Sexuality in the Speculative. Columbus: The Ohio State University Press, 2020.
Merriam-Webster. "Definition of INTELLIGENCE." Accessed August 16, 2020. https://www.merriam-webster.com/dictionary/intelligence.
MIT Technology Review. "2018 MIT Technology Review." MIT Technology Review, 2018. https://www.technologyreview.com/10-breakthrough-technologies/2018/.
Premack, David. "Human and Animal Cognition: Continuity and Discontinuity." *Proceedings of the National Academy of Sciences* 104, no. 35 (August 28, 2007): 13861–67. https://doi.org/10.1073/pnas.0706147104.

Probert, Stephen K. "Book Review: Introduction to Systems Theory." *International Journal of Systems and Society* 1, no. 1 (2014): 55–57.

Ribeiro, Marco Tulio, Sameer Singh, and Carlos Guestrin. "'Why Should I Trust You?': Explaining the Predictions of Any Classifier." *arXiv:1602.04938 [Cs, Stat]*, August 9, 2016. http://arxiv.org/abs/1602.04938.

Russakovsky, Olga, Jia Deng, Hao Su, Jonathan Krause, Sanjeev Satheesh, Sean Ma, Zhiheng Huang, et al. "ImageNet Large Scale Visual Recognition Challenge." *International Journal of Computer Vision* 115, no. 3 (December 1, 2015): 211–52. https://doi.org/10.1007/s11263-015-0816-y.

Schrittwieser, Julian, Ioannis Antonoglou, Thomas Hubert, Karen Simonyan, Laurent Sifre, Simon Schmitt, Arthur Guez, et al. "Mastering Atari, Go, Chess and Shogi by Planning with a Learned Model." *Nature* 588, no. 7839 (December 2020): 604–9. https://doi.org/10.1038/s41586-020-03051-4.

The AlphaStar Team. "AlphaStar: Grandmaster level in StarCraft II using multi-agent reinforcement learning." Deepmind, October 30, 2019. https://deepmind.com/blog/article/alphastar-mastering-real-time-strategy-game-starcraft-ii.

Trewavas, Anthony. "Aspects of Plant Intelligence." *Annals of Botany* 92, no. 1 (July 1, 2003): 1–20. https://doi.org/10.1093/aob/mcg101.

Turing, A. M. "Computing Machinery and Intelligence." *Mind, New Series* 59, no. 236 (1950): 433–60.

Two Minute Papers. "DeepMind's AlphaStar: A Grandmaster Level StarCraft 2 AI - YouTube," 2019. https://www.youtube.com/watch?v=jtlrWblOyP4.

Vinyals, Oriol, Igor Babuschkin, Wojciech M. Czarnecki, Michaël Mathieu, Andrew Dudzik, Junyoung Chung, David H. Choi, et al. "Grandmaster Level in StarCraft II Using Multi-Agent Reinforcement Learning." *Nature* 575, no. 7782 (November 2019): 350–54. https://doi.org/10.1038/s41586-019-1724-z.

Why I Draw with Robots, 2020. https://www.youtube.com/watch?v=q-zGXV4Fd1oA&t=426s.

Zhao, Jieyu, Tianlu Wang, Mark Yatskar, Vicente Ordonez, and Kai-Wei Chang. "Men Also like Shopping: Reducing Gender Bias Amplification Using Corpus-Level Constraints." arXiv Preprint arXiv:1707.09457, 2017.

PART III

The Cybernetic Environment

9

CYBERNETICS AND LANDSCAPE

From Uncertainty to Opportunity

"There is always a tension in landscape between the reality and autonomy of the nonhuman and its cultural construction, between the human impulse to wonder at the wild and the compulsion to use, manage, and control."[1] Working with living systems and materials, landscape architects must deal with unexpected outcomes, and the tension between control and uncertainty has been central to the work of landscape architecture. Meanwhile, this tension has been the basis for cybernetics, an interdisciplinary field that emerged in the United States in the 1940s when a group of postwar intellectuals, including engineers, mathematicians, anthropologists, and ecologists, converged on a new theoretical model based on systems thinking to understand control and communication between mechanical and biological systems. Landscape architecture and cybernetics should be viewed in tandem, and cross-pollination of thought could shed light on many pressing issues resulting from control and uncertainty, such as climate change.

However, it is only in recent years that scholars have begun to attend to the influence of cybernetics on landscape architecture. Landscape theorist Anita Berrizbeitia deployed concepts from second-order cybernetics such as "structural open" and "operational closure" to analyze the Downsview Park competition and illustrate how designers account for scales of undecidability with systems thinking.[2] Though her scope was limited to posthumanist Cary Wolfe's interpretation of cybernetics, Berrizbeitia's analysis showed potentially transformative outcomes in connecting the two fields.

Landscape scholar Margot Lystra builds specific connections between landscape disciplines and cybernetics within a broader conceptual framework.[3] Lystra argues that cybernetics concepts were instilled into the landscape

DOI: 10.4324/9781003320852-13

discipline via ecological science and arts in the 1960s, and Ian McHarg and Lawrence Halprin were two prominent figures among those who established the link between cybernetics and modern landscape architecture.

Ecologist G. Evelyn Hutchinson had been a member of the early Macy Conferences on Cybernetics, and he became one of the pioneers in exploring the ramifications of cybernetic principles for ecological science. Hutchinson's student H. T. Odum integrated the notion of feedback and communication into ecological science, and his work depicted ecosystems as predictable and controllable. This deterministic view was favored by environmental designers in the mid-twentieth century, including McHarg, for it provided a sense of scientific causality with which to rationalize design decisions. Lystra argues that McHarg's adoption of "entropy" was evidence that cybernetics migrated via ecology into the landscape discipline.[4] Entropy describes the tendency of a system to move toward disorder and uncertainty when there is no energy input into the system. McHarg, Odum, and Norbert Wiener shared the same sentiment toward entropy; they viewed it as a threat to the integrity of the system, whether mechanical or biological, which they sought to control. Thus, managing entropy equals reducing uncertainty and controlling the system.

Lystra argues that, in contrast to McHarg, Halprin imported cybernetic ideas via the arts, specifically musical composition and choreography. Several arts and science collaborative events in the 1960s facilitated the migration of ideas, such as "9 Evenings: Theater and Engineering" in 1966. Lystra argues that "[a]rtists used cybernetic ideas in remarkably different ways than ecologists did: they found that cybernetic notions of indeterminacy offered methods for relinquishing, rather than increasing, control over their works. Many embraced chance as a catalyst for newly participatory and open-ended compositions."[5] In this movement, artists challenged the traditional notional systems that give precise instructions to performers, and they developed an "open scoring technique" that suggests only general intentions, and thus the resultant piece is partly dependent on the performers' choices. Open scoring was then imported to the landscape discipline by Lawrence Halprin through collaborations with his wife and partner, Anna Halprin, a choreographer and dancer.

Lystra's investigation opens a new avenue of research that takes cybernetics as a framework with which to examine key issues in the landscape discipline. Nevertheless, it is rather forced and too linear to map two distinct genealogies that demonstrate how cybernetic principles were introduced into landscape architecture.

First, as recognized by Lystra, cybernetics and systems thinking were simply "in the air" by the 1960s. The over-articulation of pathways gives a false impression that the migration of knowledge is a process of one person passing it to another, as if the issue of indeterminacy originated within the field of cybernetics, was mobilized by artists, and was finally imported to the landscape discipline by Lawrence Halprin. However, by the 1960s, Halprin had in fact already been involved in multiple community engagements in his professional work, and he had already explored scoring and notational systems to cope with the issues of uncertainty in design processes. Though the issue of control and uncertainty is at the center of cybernetics, the shift in how we understand uncertainty was more of an epistemological reflection in the 1960s across disciplines rather than a development specific to the field of cybernetics and then exported to other disciplines.

Second, like many others, Lystra's analysis starts with reasoning that positions landscape architecture as a passive receiver of ideas from other disciplines, overlooking the value of landscape architecture as a model that provides a new lens with which to examine the issue of indeterminacy. The inherent disciplinary hierarchy in many landscape theorists' analysis prevents us from recognizing the value of landscape as a transformative field of research via material practice. Indeed, despite designers' borrowing of terms from other disciplines, they did not thoroughly import the concepts; terms and ideas were often interpreted very differently. Halprin's description of his scores that "communicate but do not control" apparently plays on Wiener's book, titled *Cybernetics: or Control and Communication in the Animal and the Machine*. Halprin sought to distance his approach from Wiener's cybernetics, formalizing another way to articulate control and communication deeply rooted in the tradition of landscape architecture.

Finally, it is understandable that Lystra's analysis focuses on the 1960s, when landscape architecture was going through a paradigm shift; its key figures, such as McHarg and Halprin, have left enduring legacies and impacts on the discipline. However, this limited time frame essentially ignores the dramatic development of cybernetics after the late 1960s. The advancement of second-order cybernetics in the 1970s revolutionized the field itself and inspired many transformative concepts, such as autopoiesis and emergence, that altered how scholars understand and discuss the issue of uncertainty.

Because of the field's interdisciplinary nature, concepts in cybernetics and second-order cybernetics have been instilled into diverse fields, such as sociology, computer science, systems theory, ecology, and humanities, from all of which landscape architects have drawn inspiration. It is better to recognize that cybernetics and second-order cybernetics and their concerns have formed an undertone to intellectual life in the latter half of the twentieth century. Thus, genealogizing cybernetics to landscape architecture would be less useful than juxtaposing the two to construct connections between them. By analyzing how different concepts have been used to address the tension between control and uncertainty, modern landscape architecture could become a truly transformative model in contemporary environmental practice.

In her seminal work, *How We Became Posthuman*, N. Katherine Hayles has schematized the cybernetics movement into three waves of research, where the frontier of each wave has mobilized among different fields of study.[6] Three cybernetics waves can serve as an organizational and analytical framework for juxtaposing landscape design and cybernetics and highlighting how the tension between control and uncertainty was understood and addressed in each wave. Juxtaposing landscape design with cybernetics brings forth another crucial aspect that has been overlooked within the field of cybernetics—drawing and representation, which are important visual techniques for landscape architects to identify design possibilities and opportunities. Shifts in representational techniques in the landscape discipline reflect changes in how designers interpret and manage uncertainty.

The First Wave: Homeostasis and Entropy

Based on Claude Shannon's foundational work, which formalized information and communication, and inspired by wartime mechanical control systems such as servomechanisms and artillery targeting systems, Norbert Wiener revived the concept of cybernetics to develop a general theory of control and communication among not only mechanical but also biological systems, such as humans. The first wave of research speaks to the Macy Conferences on Cybernetics (1946–1953) that brought together influential postwar intellectuals, including mathematicians Norbert Wiener, Claude Shannon, and John von Neumann; anthropologists Gregory Bateson and Margaret Mead; neurophysiologist Warren McCulloch; physicist and philosopher Heinz von Foerster; psychiatrist W. Ross Ashby, and others from diverse fields. These conferences planted the interdisciplinary seed for the field of cybernetics, which, in turn, profoundly influenced sciences and humanities in the following decades.

The first wave focused on feedback mechanisms in self-regulating systems. Engineers sought to control complex systems through feedback, following the function $y = f(y)$, with output feeding back into the system as input. There are two types of feedback mechanisms: positive and negative. In negative feedback, an increase of y at time t will decrease y in $t+1$, and vice versa; thus, the result of the function will oscillate around an equilibrium. Negative feedback, such as in thermostats, is ubiquitous. If the room temperature exceeds a threshold, the air conditioner turns on; otherwise, it is off. As a result, the room temperature will oscillate around the target temperature. In contrast, positive feedback will produce runaway behavior in which the increase (or decrease) of y at time t will increase (or decrease) y in $t+1$. While negative feedback is the key to stability, positive feedback is responsible for growth and self-organization.

First-wave cybernetics favored negative feedback over positive feedback because the former could deliver controlled equilibrium, which was often associated with complexity. For Wiener, positive feedback was regarded as "disruptive and destructive, rather than as leading to complex stable structures."[7]

Entropy was a core concept in Wiener's reasoning. First developed in the field of thermodynamics and statistical mechanics, entropy describes the tendency of a thermodynamic system to move from order and organization toward disorder and chaos, as expressed through Boltzmann's function for statistical mechanics:

$$S = kBlog\Omega$$

where Ω represents the possible state of a system in a given time, S is the entropy, and k is the Boltzmann constant. Entropy S is in a logarithmic relationship with Ω, the number of possible system states. If a system possesses only one state, then the system is predictable because the probability for this state will always be 1, and thus entropy is 0 ($log1 = 0$). Conversely, for a complex system such as the universe, the number of possible states is infinite, and entropy reaches its maximum. In such case as the universe, since the possibility of the system being in a given state is so small, it becomes almost impossible to predict the system's behavior, and thus the system reaches maximum uncertainty. Claude Shannon similarly defines information in terms of the uncertainty of a system; we can decrease information entropy (or uncertainty) by eliminating certain possible states through observation. In this definition, Shannon constructed a direct relationship between observation and uncertainty; the information we receive from an observation equals the degree to which entropy or uncertainty is reduced in this round of observation. This conceptualization also built relationships between modeling and uncertainty. One can construct more complex models to represent the phenomena through observation, thus reducing uncertainty.

The second law of thermodynamics posits that if there were no energy input, the total entropy of an isolated system would increase over time, and probability would be evenly distributed across each state of the system. For example, if we pop a balloon full of oxygen, then the oxygen molecules will eventually blend within the surrounding air; the balloon acts as energy input that holds the isolated system in a low-entropy and highly ordered state; without the balloon, the system of air molecules would move to disorder (high entropy). For this reason, the term "heat death of the universe" entails that the universe will always move toward disorder, and entropy will reach its maximum level. The influence of the second law of thermodynamics on Wiener's reasoning is apparent here:

> As entropy increases, the universe, and all closed systems in the universe, tend naturally to deteriorate and lose their distinctiveness, to move from the least to the most probable state, from a state of organization and differentiation in which distinctions and forms exist, to a state of chaos and sameness.[8]

For Wiener, to control is to reduce uncertainty, or entropy, which is the tendency to move away from an organized state; the means for control is negative feedback. "It is my thesis that the physical functioning of the living individual and the operation of some of the newer communication machines are precisely parallel in their analogous attempts to control entropy through feedback."[9] For Wiener, negentropy became a determining characteristic for life, because feedback is key to a negentropic system to "produce around it a local zone of organization in a world whose general tendency is to run down."[10] This line of reasoning finally connected the idea of entropy to homeostasis, as argued by Wiener that "[t]he process by which we living beings resist the general stream of corruption and decay is known as *homeostasis*."[11]

Entropy had a negative connotation in first-wave cybernetics, as did uncertainty; they existed on the same line with corruption, deterioration, and decay. Homeostasis was preferred, and control systems were a means to achieve observed stability. This reasoning could be easily extended to framing wild nature versus controlled landscapes. Wild nature represents uncertainty with high entropy, and technology is the means to bring order to wildness, producing homeostatic and habitable landscapes. In this view, landscape practices become a negentropic mechanism that fights against decay and deterioration. From this vantage, Ian McHarg's ecologically inspired design methodology could be framed as a version of first-order cybernetics. McHarg used the concept of entropy extensively throughout his writing and contrasted entropy with ecological fitness.[12] The ecological science to which McHarg was exposed in the mid-twentieth century was dominated by Clements community ecology, which emphasizes ecological successions toward a climax community that is stable and desired. McHarg also emphasized succession and climax species. His design principles imply that environmental factors are determined; if we "design with nature," the ecosystem will always follow a predetermined and predictable path toward the climax community, a homeostatic equilibrium.

Aside from the underlying ecological framework that favors stability over uncertainty, McHarg's design methodology also ensured predictability in landscape design and "embraced the measurable and controllable aspects of landscape systems."[12] He implemented a series of analytical strategies to reduce uncertainty in every aspect of landscape design. One of the most famous McHargian methods is the map overlay technique, known as "the ecological inventory" or "the layer cake." This inventory allowed designers to observe landscape systems through a set of fixed lenses always in the same order: climate, geology, hydrology, soils, vegetation, and wildlife.[13] According to McHarg, this order was meant as a chain of deterministic reasoning, with the former variables predicting the latter. By knowing climate conditions and geology, one can comprehend the hydrological patterns of a place; then, with those three variables, one can predict patterns of soil and vegetation types.[14]

These landscape features will be drawn on individual maps and then overlaid atop each other to produce a composite map for suitability analysis. If one designed the landscape based on the suitability analysis, then the future of the landscape would be predictable, and thus uncertainty would be reduced.

Despite its inherent ecological determinism, the inventory reduced uncertainty in the creative process as well. McHarg essentially standardized the design process by attributing creative agency to "objective" and "scientific" procedures so that the same protocols could be practiced repeatedly on different sites and by different designers. Design outcomes depended solely on the different inputs fed into his "landscape designing machine," and the outcomes were comparable. Thus "design with nature" constructed a model of "nature" and allowed the initial variables, decided by the characteristics of different site conditions, to determine the designed landscape. The map overlay technique systematically reduced uncertainty by inventing an autonomous protocol independent of designers. Accompanying the systemic design method was a type of drawing technique that tried to eliminate ambiguity in representation. In the overlay maps, different landscape features were drawn with distinct boundaries without overlaps, and "the clarity of imagery implied that their information was absolute, reinforcing a notion that the viewer could be certain about the veracity of the visual data they observed."[15]

McHargian ecologically inspired landscape design managed uncertainty, from underpinning conceptual framework to design methodology and representational techniques. His legacy was not limited to the landscape profession. McHarg's map overlay technique directly influenced the later development of geographic information systems (GIS) and the founding of the Environmental Systems Research Institute (ESRI), which later became the largest geographic research and service provider in North America, playing an important role in standardizing organization and representation of geospatial information. The GIS is more than a piece of software; it is a system through which the environment can be reduced to spatial information that exhibits specific structures and organizations. Although GIS plays a vital role in today's environmental management discourse, which emphasizes adaptation and flexibility, the way in which the environment is structured as spatial information remains within the ecological paradigm that promotes stability through managing entropy and uncertainty.

First-order cybernetics is a model-making paradigm; it entails a sort of epistemological realism, where the object of study is independent of the human mind. With this conceptualization, "all our knowledge is mediated by our simplified representations—or models—of them, which necessarily ignore those aspects of the system which are irrelevant to the purposes for which the model is constructed."[16] In environmental practices undergirded by first-order cybernetics, there will always be models of the environment on which our decisions are made. In a way, the McHargian ecological inventory is a means to construct a model of the environment that predicts how the landscape could evolve and how design intervention could impact that evolution. Similarly, many present-day environmental practices, including adaptive management, rely on complex digital models to predict and simulate control strategies and evaluate their impacts, such as the South Florida Water Management Model used to devise watershed management strategies. In this conceptualization, uncertainty becomes the gap between the model and the modeled environment; uncertainty may be reduced via more complex models.

The Second Wave: The Irony of Relinquishing Control

Despite Wiener's original intention to develop a general control and communication theory across different systems, cybernetic principles quickly found their direct applications in the design of mechanical systems. The first wave of cybernetics gave rise to robotics, system engineering, computer science, artificial intelligence, and many other application-based fields, quickly drawing scholars' attention away from theory-based inquiries. In the meantime, cyberneticians, including Gregory Bateson, Margaret Mead, and Heinz von Foerster, felt the need to update cybernetics to focus on the issue of reflexivity and problematize the innocent observers of the first wave.

This concern over the observer mirrored a broader societal reflection on authorship, objectivity, and scientific truth in the twentieth century. In science, Thomas Kuhn's *The Structure of Scientific Revolutions* (1962) found its popularity beyond academics; so did the term "paradigm shift." Kuhn explored an alternative to the "great man" history of the scientific revolution, arguing that scientific knowledge was constructed by scientific communities based on their shared consensus regarding the underlying assumptions, techniques, and values reached by their members. In other words, individual excellence was questioned, and scientists and their unconscious minds became subjects of study.

Similarly, problematized authorship was a key issue in early twentieth-century works of art. Marcel Duchamp's readymade sculpture "Fountain" was one of the earliest avant-garde landmarks to challenge the role of artists in art production. A standard urinal, signed and dated "R. Mutt 1917" by Duchamp, was submitted to a jury-free exhibition of the Society of Independent Artists in April 1917. However, the Society's board of directors excluded Duchamp's submission based on the alleged indecency of the work. The tension between artists, art authority, and art critics posed questions about the authenticity of the works of art, and problematized authorship became one of the core issues in the later conceptual art movement. Conceptual artist Sol LeWitt argued that the idea was the most important aspect of art production, and the idea was the machine that produced art. In 1968, LeWitt began his wall drawing series, for which he only gave instructions on how to make drawings on the walls. These instructions were "scores" that choreographed his installation crews to produce art.

Similar concerns can be found in humanities as well. In "The Death of the Author" (1967), Roland Barthes questioned the role of authorship and argued that an author was not the creator of meaning in the text, but merely someone who meshed different texts. Once the author was gone, the text was truly liberated and overflowed with possible interpretations.[17] In the same vein, Michel Foucault questioned the issue of authorship with increased sophistication. More than activating readers by removing authors, authorship for Foucault can be understood through author function, which speaks to the need to pin a discourse to a person as a regulator of meaning. The author becomes an "ideological figure" implicated in our understanding of discourses.[18]

Many regarded Barthes and Foucault as key figures in the post-structuralism movement. Drawing ideas from linguist Ferdinand de Saussure's structural linguistics, structuralists deployed language as a model to analyze cultural products and society itself, prioritizing systematic inquiry to interpret their underlying structures. Whereas post-structuralists such as Barthes and Foucault have challenged the unquestioned linguistic model and have accepted that language is a flawed means of communication. Thus, reading through a text and deciphering the author's original intention is impossible and meaningless. Most importantly, for post-structuralists, this type of analysis, which focuses on interpreting the texts' underlying structure, ignores its rich possibilities. Thus, post-structuralism encourages us to reject any final meaning of a text and instead embrace the multiple meanings of cultural products. Post-structuralism began with a sense of reflexivity and problematized authorship but later found a broader implication in many fields, such as Jacque Derrida's deconstruction. In a way, post-structuralists embraced the uncertainty of the flawed linguistic system, and used it to their advantage to deconstruct the underlying binaries in the structures that give rise to cultural phenomena.

These examples demonstrate that in the mid-twentieth century, cybernetics was not the only field grappling with the issue of reflexivity. As argued by Cary Wolfe, at least two genealogies can be traced to the rise of posthumanism, one on the line of Foucault and post-structuralism and another on the route of second-order cybernetics.[19] From this perspective, second-order cybernetics essentially provides a systemic account for the issues of problematized authorship, presenting another model for posthumanist epistemology based on systems thinking.

Early cyberneticians such as Gregory Bateson realized that the issue of reflexivity is more than subjectivity in modeling but is, instead, an issue in the epistemological framework. Bateson predicted that "the problems posed by including the observer could be addressed only if a substantial reworking of realist epistemology was undertaken."[20] To resolve the issue of reflexivity requires a fundamental shift in the epistemological framework, and, as we have seen, constructivism was already discussed by the late 1960s. Cyberneticians were ready to embrace a more radical approach to the issue of reflexivity, one which was not properly addressed in the Macy Conferences era. In 1967, Margaret Mead, one of the key participants in the Macy Conferences, addressed the American Society for Cybernetics regarding the need for a recursive application of cybernetics on itself, treating the observer as a cybernetic system constructing models of another cybernetic system.

Early cyberneticians, including Mead, Bateson, and Heinz von Foerster, proposed a model of second-order cybernetics by drawing a larger box around the original feedback diagram (Figure 9.1). Nevertheless, this model was an intuitive leap until Chilean biologist Humberto Maturana's research on the frog's cortex provided empirical evidence for second-order cybernetics. According to Maturana, the frog's eye does not send a whole data package of the outside environment to the brain to interpret; instead, the environment triggers only a set of nervous system operations between the eye and the brain that produce a reality for the frog itself.[21] This finding inspired Maturana and Francisco Varela to coin the term "autopoiesis (self-production)" to describe this phenomenon. "The living organization is a circular organization which secures the production or maintenance of the components that specify it in such a manner that the product of their functioning is the very same organization that produces them."[22] Autopoiesis is essentially a version of radical constructivism claiming that knowledge is constructed through system operations to maintain the organization that constructs that knowledge.

In this highly reflexive framework, uncertainty becomes a notion which human systems invent to distinguish the conditions of known and unknown; uncertainty also plays a role in how we describe learning and time. Based on autopoiesis theory, a system, such as an observer, can only behave in the present by interacting with the descriptions that the system generates at

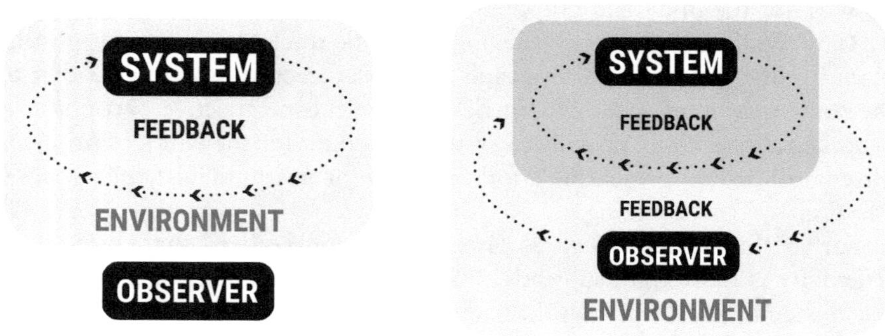

FIGURE 9.1 Feedback diagram. Illustration by the author.

this moment. Paradoxically, the observer could interact with a description generated in the past as if it were in the present. To resolve this paradox, the observer thus needs a notion of time—past, present, and future—to expand the domain of interactions. In autopoiesis, it is not because the notion of the future exists in the first place that uncertainty is on the horizon. Instead, it is because humans need a concomitant emotional state of anxiety when the observer interacts with a description for the first time, and the notion of uncertainty can provide that state of anxiety to maintain the system's organization. "Any experience without anxiety [or uncertainty] can be described as known, and thus serve as a basis for the functional notion of time."[23] In a sense, autopoiesis flips our conventional understanding of the relationship between uncertainty and future. Uncertainty becomes a notion to describe the unknown; consequently, reducing uncertainty becomes a description of learning.

Despite criticism of its self-referential language and potential solipsism, autopoiesis essentially problematized the observer and challenged the realist epistemology and objectivity in early cybernetics research. Radical constructivism found implications, especially in the posthumanist arguments. N. Katherine Hayles drew inspiration from autopoiesis and argued that those qualities, such as consciousness, intelligence, and free will, that helped those with the privilege to conceptualize themselves as liberal human subjects, were, in fact, epiphenomena of the system operation. Autopoietic theory "sees thinking as a secondary effect that arises when an autopoietic entity interacts with its own representations. Self-consciousness, a subset of thinking, is relegated to a purely linguistic effect."[24]

If we view the second wave of cybernetics as wrestling with authorship and challenging observers' authority, then we can identify many landscape counterparts in the second half of the twentieth century that questioned designers' intentions and authorship. Lawrence Halprin's score is one of them. Though Margot Lystra acutely defined Halprin's scoring method as a systemic approach informed by cybernetics, a caveat is that Halprin should be juxtaposed with specifically second-order cybernetics. In an interview, Halprin stated that his notion of scores emerged from a realization that if people were excluded from the design process, they would grow angry. Thus, he invented a system whereby all felt that they were shareholders, not only as clients but also as participants in the creative process.[25] Halprin offloads his authorial control to a system of scores, and the design outcome seems to become a byproduct of the system operation, independent of the designer's excellence. However, just as second-order cybernetics' use of a bigger box around the original system diagram created a meta-observer who hides behind a second-order cybernetics diagram, Halprin's scores essentially elevated designers to another plane of operation, to perform a sort of "distanced authorship."

Urbanist Charles Waldheim introduced the term "distanced authorship" when characterizing a body of landscape projects from the late 1980s to the early 2000s. "A variety of contemporary landscape practices evidently employ techniques of problematized authorship and contemporary discourse around landscape and urbanism is awash with claims of indeterminacy, open-endedness, self-regulation, and post-modern ecological models of autonomous emergence."[26] Waldheim argues for three categories of autonomous systems—computer algorithms (technical), activated users (cultural), and ecosystems. Yet, Waldheim's categorizations miss an underlying epistemological shift in landscape design. We should ask a different question: what is being designed? Since the 1980s, designers have grown interested in designing landscape systems rather than objects, programs, or forms. It is not that forms and programs are unnecessary, but they are no longer the fundamental components in how design projects are articulated and justified. Landscapes are envisaged as systems so that forms and programs become epiphenomena and concomitant effects in system operation. They matter only with regard to their functions and performances as components in the landscape systems. In a way, "distanced authorship" performs a type of second-order design: designing a system that designs. In this way, indeterminacy is achieved by instrumentalizing the self-production of the landscape systems that are imagined to be autopoietic living systems, and that exhibit a certain level of autonomy.

The quality of self-production can be illustrated through examining in tandem two competition projects by OMA: the first is the entry for the Parc de la Villette competition in 1982, and the second is the entry "Tree City" for the Downsview Park competition in 1999. In the Parc de la Villette, OMA's entry is considered a park of "culture" rather than "nature." The site was divided into a series of stripes of autonomous programs. The programs in different stripes "pollute" each other, making the park unstable and out of the designers' intended control.[27] These stripes are unstable because users, rather than authors, define their programs, whose characteristics are thus contingent on cultural practices over time.

Similarly, in the Downsview competition, OMA proposed a network of 1,000 paths, with vegetal clusters separated by vast undesignated areas, as a spatial framework to account for the park's future development. Many argue that OMA's "Tree City" exhibits some level of uncertainty, thus is open-ended. However, this flexibility is based on the self-production of the urban park as an autopoietic system. The project description resonates with autopoiesis theory, where the living systems operate to reproduce their organization:

> Tree City treats the park as if it is an adult soon capable of sustaining itself rather than a child in need of eternal care...We propose that capital

generated from the park's appreciated land value be spent to manage the park's infrastructure and to support future development in an evolving cycle of implantation and speculation.[28]

Another way to understand self-produced flexibility is through the redefined system–environment relationship in autopoiesis theory. Since an autopoietic system interacts only with a representation constructed by itself, the outside environment is no longer accessible by the system. *Structural coupling* was introduced as a concept to describe interactions between systems. One of the contributions of autopoiesis theory is to distinguish structure from the organization—"the organization of a system as the set of relations between its components that define it as a system of a particular class, is a subset of the relations included in its structure."[29] The structure of a system can change without affecting its identity, as long as the organization of the system remains invariant. Structurally, a system is coupled with its environment and other systems; that is, there are pathways and relations between the system and its environment through which the system can exchange information, material, and energy flows. These pathways depend on the system's structure—how the system is wired concretely. When the system operates, it uses flows of information, material, and energy to reproduce its own organization so that it retains its identity. For this reason, autopoietic systems are said to be "structurally open" but "operationally closed."

In both of OMA's proposals, flexibility exists on the structural level. The stripes in the Parc de la Villette are structural so that they can be highly flexible—the forms, programs, functions, and performance within these stripes could be of any sort. As long as they exist within the park's organizational framework, the park's identity will not be jeopardized by changes on the structural level. Similarly, in "Tree City," the paths and vegetal clusters are not organizational components, so "clusters can acquire the form of precise, static, round shapes without the scheme losing any of its flexibility, complexity, or openness...What is inside the 'dots' responds to other criteria, independent from the park/environment organization."[30] The structure/organization difference thus creates two types of relationships between the site and its environment—structural and organizational. The park is structurally coupled with other urban systems so that it can be structurally flexible toward contingencies posed by these systems. Yet, the park is operationally closed to these external systems because it must retain the identity of a suburban Toronto park. As described by the designers, "[v]egetal clusters rather than new building complexes will provide the site's identity...Tree City assumes the park's suburban context to be its virtue. The locale offers an ideal opportunity to explore the unrealized promises of low-density metropolitan life."[31]

Once we take inspiration from autopoiesis and analyze "distanced authorship," we can find that forms, programs, and landscape objects, once the core of landscape practices, now become structures that are separated from the organizations of landscape projects. Anita Berrizbietia thus praised second-order cybernetics and autopoiesis framework, because through the structure/organization difference,

> we can conceptualize landscapes where there is space and time for process to unfold and for stable meanings to come forth… We have always known landscapes to perform multiple programmatic, functional, and metaphorical roles simultaneously, shifting from one system of signification to another, but only recently have we been able to articulate this complexity under a single framework.[32]

Though the ideas of flexibility and uncertainty might have been novel in the late 1990s and early 2000s, after nearly 20 years of development, these ideas have been tested in landscape practices. This new condition serves as a basis for us to reflect on autopoiesis and "strategies of indeterminacy" and ask whether they are truly indeterminate.

Structure/organization creates another level of stability, paradoxically achieved by amplifying structural flexibility. Is OMA's proposal flexible? On a structural level, yes. Is it truly open-ended and indeterminate? No, because the park is envisaged in such a way that information and material flowing into it are used to reproduce its organization and maintain its fixed identity as a suburban park. The paradox here reflects a shortcoming in the autopoiesis theory. As pointed out by Hayles, even though the implication of autopoiesis theory challenges observers' authorial control, it is still on the line of liberal humanism that re-inscribes a set of humanist values, including homeostasis.[33] The second wave of cybernetics raised questions about the observers but less about the conception of homeostasis that was central in the Macy Conferences era; homeostasis in first-order cybernetics persisted in the autopoiesis theory. Maturana and Varela stated clearly that "living systems are a subclass of the class of circular and homeostatic systems."[34] Autopoietic systems operate to maintain their equilibrium, and there remains little room to consider change and emergent behaviors. Structure/organization may resolve certain issues regarding system structural flexibility, but it becomes difficult to explain evolution and change; once the organization changes, the system becomes a different class. However, if a system operates to maintain the circularity that secures its organization, how could a system stop being one thing and become another, and how could evolution occur?

The individuation process is concurrent with the attribution of agency, and so is operational closure. Thus, autopoiesis should not prevent us from understanding the tendency for systems to assimilate, aggregate, and mobilize in a heterogeneous field. Drawing inspiration from ecological science, organizational closure at its best should be considered a local equilibrium, and autopoietic systems can move into another temporary stability. However, both of OMA's projects represent a strain of environmental practices that claim to be indeterminate on a structural level yet, in reality, re-inscribe the notions of stability, determinism, and homeostasis via organizational control. Not all "strategies of indeterminacy" are truly indeterminate. Ironically, autopoiesis theory results in another form of control regime that can only be revealed within a second-order cybernetics framework; it reinforces organizational control by relinquishing structural control.

The Third Wave: Emergence and Posthumanist Ecology

According to N. Katherine Hayles, "emergence" distinguishes the third wave of cybernetics research from the second wave. It became apparent that self-production in autopoiesis was insufficient to consider runaway behaviors and positive feedback loops that move the systems from equilibrium and toward unexpected directions. Hayles argues that the frontier of cybernetics research moved into the field of artificial life around the 1990s. One thread of autopoiesis arguments is the attempt to redefine life—all living systems are autopoietic systems, and any system, if it is autopoietic, can be said to be living.[35] Thus the third wave of cybernetics may be understood as an implication of this argument in the field of artificial life, drawing analogies between computer simulations and the life evolution.

In the 1990s, evolutionary biologist Thomas S. Ray developed a computer simulation called Tierra, in which programs were given simple rules to replicate themselves by copying bits in computer memory.[36] When the machine copied the code, the programs would flip the bits (1 and 0) so there were opportunities for mutations to give birth to a new "digital species." Tierra programs could evolve into ecosystems with various digital species and communities by repeating simple rules. The results might be completely different if the initial variables were slightly changed. These computer programs compete for CPU time and space in the computer memory; they are digital species in a computational universe. Based on the empirical evidence, Hayles made the stretch to argue that human consciousness, intelligence, thinking, and the idea of free will, as well as humanist concepts used to define human subjects, can be understood as emergent behaviors of random interactions of distributed systems.

Autopoiesis in the second wave of research has evolved into a posthumanist epistemology, which offers new ways to frame concepts such as agency, intelligence, evolution, and emergence. Within a posthumanist and emergence framework, autopoietic systems could employ their flexibility to expand and construct new pathways and form structural couplings with other systems. Emergence describes the process of forming these pathways between different systems. Only through this posthumanist framework can we begin to make sense of the value of a body of landscape practices since the 2000s that emphasize emergence and posthuman ecology.

James Corner, Stan Allen, and Nina-Marie Lister's entry "Emergent Ecologies" for the Downsview Park competition might be aligned with the third wave of cybernetics, which considers emergence. This entry is in a different paradigm from the other four finalists, including the above OMA's winning proposal "Tree City." For this reason, "Emergent Ecologies" holds a high value in the landscape disciplines, even though it lost the competition. The Downsview competition brief specifically asked the teams to consider a phasing strategy for developing the park over 15 years, in five-year increments, and accommodating future changes. The five finalists all addressed this request in one way or another. The phasing strategy distinguishes "Emergent Ecologies" from the other four finalists because it considers not only the time involved in developing the park but also how the park's development will continue in the future, after construction. For the other teams, including OMA, designing a park is the ultimate goal; there will always be an endpoint when the park is considered "built," and "phasing" is limited to a strategy of spreading out its construction over 15 years. This mindset might be due to the fact that these teams were led primarily by architects, with specific images of "landscape" and "urban park" in mind.

Landscape architect and theorist Kristina Hill was highly critical of OMA's proposal, and she pointed out that OMA's team "proposed the pastoral forms of traditional urban parks, but used a language of mimicry to describe them, calling the park a 'Tree City' that celebrates (and in formal terms, imitates) its suburban context of lanes and clusters of trees."[37] Berrizbeitia has issued similar criticism:

> the OMA scheme falls short of its promise…instead of seeing the broad array of possibilities afforded by their process, we see nature and urban life reduced to one vision, that of logolike landscapes frozen as stage sets of human and vegetal performances.[38]

For landscape architects, the question of "what landscape is" has always been at the core of the profession, and practitioners have pushed the boundary of landscape practices over the course of the twentieth century. By the late 1990s, landscape architects were accustomed to understanding landscapes through processes, and recognizing landscape design as a temporary intervention on a site with a history and future of its own. This realization led the "Emergent Ecologies" team to approach the Downsview site differently:

> We propose a carefully gauged framework…this framework will be mostly constructed (or 'seeded') within the first five years, its 'lifeline' drawing energy, life, matter and activity across the site. The park's identity will

subsequently evolve and be re-shaped as users inscribe their own traces into its various surfaces and pathways over time.[39]

"Emergent Ecologies" interpreted the phasing requirement with a landscape sensibility. The designers did not treat park design as a kind of courageous adventure that imposes a "master plan"—an image of the pastoral suburban park found in major cities by the end of the twentieth century—on the site of Downsview in a 15-year timeline. Instead, the proposal treasured the first five-year window as a chance to intervene in an ongoing ecological process on the site of Downsview itself. Without designers, the Downsview site would still be a complex system that evolved and unfolded; the role of designers was to plant a seed that might or might not guide the Downsview site to assume a different trajectory into the future. As the team stated, "*We do not determine or predict outcomes;* we simply guide or steer flows of matter and information."[40] From this perspective, "Emergent Ecologies" possesses the highest indeterminacy among the five finalists, leaving the design truly open-ended. After the 15-year phasing program, "Emergent Ecologies" also considered the site's long-term potential. In other words, building a conceivable and stable suburban park was not the goal of its proposal; instead, the aim was to prepare the Downsview site for a territory full of potential. Its proposal leaves space to imagine a wild, vibrant place entirely different from a pastoral park. This sort of risk-taking spirit perhaps intimidated the Toronto jury, propelling it away from "Emergent Ecologies" in favor of OMA's "Tree City" as a less threatening proposal.

We can observe a sort of posthumanist ecological thinking that underpins the ethical framework of the proposal for "Emergent Ecologies." Even though all competition finalists addressed ecological dynamics in their own responses, only "Emergent Ecologies" recognized that ecological processes do not occur solely for human use. In other words, OMA's proposal is trapped in a humanist ecological framework that regards "nature" as resources that sustain a "low-density metropolitan life." In contrast, "Emergent Ecologies" could be understood as a posthumanist project that specifically renders a future park that can sustain more-than-human life forms. For the other entries, humans were still envisioned as a force outside the site's ecology. Yet, "Emergent Ecologies" regarded designers' interventions as an indispensable part of the site's ecosystem dynamics, so design strategies could influence its future evolution.

This level of realization can be explained through a posthumanist ecology that requires designers to cultivate a deep understanding of ecological science. As explained by Kristina Hill, ecological science since the late twentieth century has expanded its considerations in three different aspects. The first is the temporal expansion through which ecologists are willing to consider ecosystem dynamics on a geological time scale. The second is the expansion in spatial scale, where local ecological flows are part of broader ecosystem dynamics. The third is a more dynamic view of physical landscapes, as constantly shifting mosaics with different ecosystem types constantly emerging and disappearing.[41] When considered within a posthumanist framework, these updated understandings of ecology result in a series of implications, in terms of design activities. If we take the Downsview site as an example, the site's current condition and the proposed suburban park only matter on a human timescale. When considered on a geological timescale, any intervention becomes a transient phenomenon, like a drop of water in a river of evolving ecologies.

The posthumanist ecological framework also possesses ethical implications. The site is being made and remade with more-than-human species, among which a human is minuscule. From this perspective, design activity is no longer about envisioning a "master plan" and expecting changes, but instead about inserting designers' current limited understandings of the site into an ongoing process of co-production and co-evolution among more-than-human lives. With this reconfiguration of design comes an updated sense of responsibility. In a way, the designer's claim for "Emergent Ecologies," that "we do not determine or predict outcomes," is not about escape from responsibility. On the contrary, it suggests assuming more responsibility, by recognizing the limitations of designers' current ethical and epistemological frameworks that undergird and limit the proposed strategies, and expressing a willingness to step aside and give way to future interpretations of the site. From this vantage, beyond the open-ended strategies, the flexibility of "Emergent Ecologies" also lies in its ability to account for changes in ethical and epistemological considerations of future designers who will wish to insert their understandings of the site into the process of co-evolution.

Like the McHargian map overlay technique that uses defined boundaries to re-inscribe a sense of stability and reduce uncertainty, representation techniques in this wave of landscape design highlight the considerations of emergence and increased complexity over time. Representational techniques can be better illustrated by including another important landscape project of the 2000s, performed by Field Operations: the Freshkills Park design competition. The winning team proposed transforming a previous landfill on Staten Island into parkland. This time, the jury favored the proposal by Field Operations. The designers proposed a timeline to evolve the landfill into a park over 30 years through ecological emergence. Freshkills and Downsview should be viewed as seriation because they share not only the underlying design framework concerning emergent ecologies but also a similar representational technique. Unlike most "master planning" projects that feature a top-down view plan and perspectival rendering that freeze time, Freshkills and Downsview feature a series of site maps in a sequence, displaying increased ecological patches, habitats, and human activities over time.

Accompanying the map sequences are a series of abstract timeline diagrams portraying the emergence of wild animals and plant species, indicating enhanced biodiversity over time. The use of timelines de-emphasizes a single view of the project through one master plan; both projects represent the landscape as a process of emergence rather than a static conglomeration of objects, forms, and functions. Unlike the McHargian map overlays technique that reduces ambiguity, Field Operations' diagrams are abstract and open to interpretations. Especially in the case of Freshkills, the timelines consist of abstract dots rather than concrete forms. Yet it matters little what the dots represent; what matters are the increased quantity, type, and colors of these dots over time. These diagrams invite different interpretations about emergence and increased complexity.

In both their representational technique and underlying ecological framework, the Downsview and Freshkills entries led by Field Operations became early-2000s exemplars for a new paradigm of landscape design deeply rooted in cybernetics. Over the past two decades of practice and theory-development, the discipline of landscape architecture has developed a design framework in which strategies are articulated through process-based operations that aim to intervene in socio-cultural, technological, and ecological processes. The designs are said to be open-ended with emergent qualities unfolding over time. Most importantly, uncertainty is no longer viewed as a negative concept; instead, it becomes a source of emergence. At both Downsview and Freshkills, the sites possess rich possibilities to take on different trajectories *because* the future is uncertain.

Three Paradigms of the Cybernetic Environment

By juxtaposing the development of landscape practices with the evolution of cybernetics, we can see that the key concerns in each wave of cybernetics mirrored how designers consider the issue of control and uncertainty in landscape practices since the mid-twentieth century, when designers began to instill cybernetic principles into their design frameworks. These changes in considerations can be described by three paradigms of environmental practices undergirded by cybernetics (Figure 9.2).

The McHargian ecological design framework operates in first-order cybernetics. This type of environmental strategy seeks to maintain a homeostatic system by reducing entropy or uncertainty, which represents the model/environment differences. Its consequent reasoning is that the gap between the model and the modeled environment can be reduced by constructing more complex models through iterative system observation and monitoring, and thus uncertainty can be regulated. Several versions of adaptive management, which rely on model-making and digital simulations for decision-making, rest within this category.

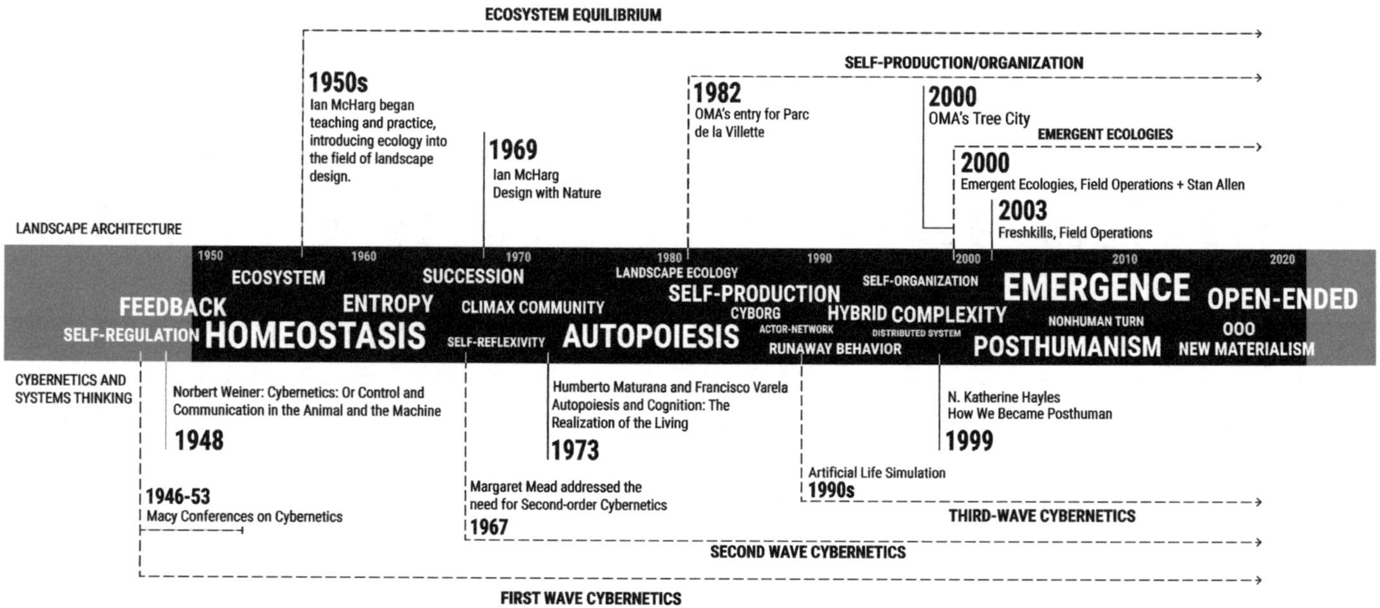

FIGURE 9.2 Landscape design and cybernetics. Illustration by the author.

The second paradigm speaks to Halprin's score and OMA's two park competitions in the 1980s and 1990s. Through problematizing authorship in design processes and emphasizing organization/structure differences, this type of practice articulates an overall homeostatic framework on the organizational level, within which flexibility is produced by distributed authors/users on the structural level of the system. Uncertainty is instrumentalized as material and information flow which park systems can use to reproduce their organization, thus maintaining their fixed identities. A new type of control regime is established on the organizational level by relinquishing control on the structural level.

The third paradigm includes the landscape practices exemplified by Downsview Park and the Freshkills entries led by Field Operations. This paradigm reflects the notion of homeostasis and favors emergent behaviors and posthumanist ecology.

Even though the three paradigms emerged in chronological order, the development cannot be described as two paradigm shifts because the last paradigm has not entirely overthrown and replaced its predecessors. For example, McHargian map overlay is still prevalent in today's landscape and ecological planning, and ecological fitness, in many cases, is still associated with stability and equilibrium.

Similarly, the second paradigm—structural flexibility within a homeostatic framework—has also been observed in recent environmental practices. Many contemporary urban design projects rely on programming and events as strategies to revitalize designed public spaces. For example, the Bentway, designed by Public Work, is an urban landscape project located beneath the Gardiner Expressway in Toronto. The project relies on the strategy of intentional programming to revitalize an undesired space under a mono-functional transportation infrastructure. The design was conceptualized around the expressway's columns, called "bents," that could form 55 "civic rooms" to be used either as a whole, as in a winter skating trail, or individually, with different events from farmers' markets to contemplative spaces. The designers consulted neighboring residents and potential users to develop a range of potential events which might be hosted in the Bentway, and the designers needed only to provide basic infrastructures to support these activities. The Bentway Conservancy was established as an agency to maintain the public space and, most importantly, organize and program events and activities throughout the year. The Bentway project uses the same strategy as in OMA's Park de la Villette design, and cultural practices and uncertainties are conceptualized as desired qualities under a fixed-system organization and identity that fit within a linear urban park typology.

After Hurricane Sandy hit the New York City region in 2012, the Rebuild by Design competition called for projects to address the issue of climate resilience through landscape strategies. Led by Bjarke Ingels Group (BIG),

The BIG U proposal was one of the finalists. It proposed a protective system in Lower Manhattan to run from West 57th Street to the southern tip of Manhattan Island and then back up to East 42nd Street, forming a U-shape to shield Lower Manhattan. The U-shape comprises a series of flood infrastructures which could be used as public spaces for all sorts of activities, from jogging to practicing tai chi, as rendered by the architects. The Rebuild by Design competition specifically requested the teams to conduct community engagement in the design process. In a way, community feedback could be understood as uncertainty that undermines designers' intentions. From this perspective, the flexibility of the BIG U manifests in its ability to accommodate various inputs from stakeholders and community members and provide multitudes of activities to meet their needs. However, that flexibility is based on an aggressive and dominating organizational framework—the U-shaped protective system. In other words, in the BIG U, we observe a tactic similar to one practiced by OMA in the Park de la Villette and the Downsview competition: because the suggested programs and activities rest on the structural level, they can be of any type as long as the overall organization of the system remains invariant. Since the U-shape remains unchanged, it matters very little what sorts of activities can occur and in which forms the infrastructures are built. Community feedback poses uncertainties on the structural level, and they are instrumentalized in maintaining the identity of the BIG U through autopoietic self-production. In a way, community members are encouraged to accept the BIG U as a given constant, and thus community engagement should never pose any genuine challenge to those characteristics that give the project its identity; on the contrary, the more feedback generated, the more flexible the BIG U appears to be to accommodate uncertainties.

This can be observed in recent years, when the BIG U was implemented in the planning and design of several sections of flood infrastructure in Lower Manhattan, named the East Side Coastal Resilience (ESCR), Brooklyn Bridge Montgomery Coastal Resilience (BMCR), Lower Manhattan Coastal Resilience (LMCR), and Battery Park City Resilience (BPCR) Projects. The ESCR created the most controversy among these implementations when the demolishment of the existing East River Park began in December 2021 despite protests and objections from an activist group called "East River Park ACTION," which blamed the plan for being "destructive." The ACTION group critiqued the plan, calling it "ugly and cruel" and "environmentally destructive," yet the activists could not fundamentally challenge it because their objections remained structural rather than organizational. The aggressive and dominating U-shaped protective system remains an organizational framework.

The ESCR proposes deconstructing the existing East River Park to build a new park on an elevated flood infrastructure. The rendering of the park shows yet another conventional pictorial urban park programmed with fields and playgrounds—an elevated urban park on a flood wall is the least innovative approach for coastal adaptation. The BIG U fails to offer any room to imagine an alternative coastal landscape that sustains a new urban coastal ecosystem for a new type of lifestyle. The organization of the BIG U and its implementations is a homeostatic framework that seeks to maintain the status quo of Lower Manhattan—including its high-density financial center, which is, ironically, a fundamental part of the world's capitalist system which many would argue is responsible for climate change-induced extreme weather such as Hurricane Sandy. From this vantage, the Rebuild by Design competition and the resilient mindset create an ultimate organizational framework that promotes homeostasis and stability over change and uncertainty.

Notes

1 Spirn, "Constructing Nature: The Legacy of Frederick Law Olmsted," 113.
2 Berrizbeitia, "Scales of Undecidability."
3 Lystra, "McHarg's Entropy, Halprin's Chance."
4 Lystra.
5 Lystra, 77.
6 Hayles, *How We Became Posthuman*.
7 Heims, "Introduction to 'The Human Use of Human Beings,'" xv.
8 Wiener, *The Human Use Of Human Beings*, 12.
9 Wiener, 26.
10 Wiener, 34.
11 Wiener, 95.
12 Lystra, "McHarg's Entropy, Halprin's Chance." 75.
13 Spirn, "Ian McHarg, Landscape Architecture, and Environmentalism: Ideas and Methods in Context."
14 Spirn, 107.
15 Lystra, "McHarg's Entropy, Halprin's Chance," 76.
16 Heylighen and Joslyn, "Cybernetics and Second-Order Cybernetics," 3.
17 Barthes, "The Death of the Author."
18 Foucault, "Authorship."
19 Wolfe, What Is Posthumanism?
20 Hayles, *How We Became Posthuman*, 132.
21 Lettvin et al., "What the Frog's Eye Tells the Frog's Brain."
22 Maturana and Varela, *Autopoiesis and Cognition*, 48.
23 Maturana and Varela, 24.
24 Hayles, *How We Became Posthuman*, 149.
25 Halprin, *Lawrence Halprin on Design: RSVP Cycles*.
26 Waldheim, "Strategies of Indeterminacy in Recent Landscape Practice," 82.
27 OMA, "Parc de La Villette." https://www.oma.com/projects/parc-de-la-villette. Last accessed: April 29, 2024.
28 OMA, "Downsview Park." https://www.oma.com/projects/downsview-park. Last accessed: April 29, 2024.
29 Maturana and Varela, *Autopoiesis and Cognition*, xx.
30 Berrizbeitia, "Scales of Undecidability," 124.
31 OMA, "Downsview Park."
32 Berrizbeitia, "Scales of Undecidability," 124–25.
33 Hayles, *How We Became Posthuman*, 142.

34 Maturana and Varela, *Autopoiesis and Cognition*, 48.
35 Maturana and Varela, *Autopoiesis and Cognition.*
36 Ray, "An Approach to the Synthesis of Life."
37 Hill, "Urban Ecologies: Biodiversity and Urban Design," 99.
38 Berrizbeitia, "Scales of Undecidability," 125.
39 Czerniak, *Downsview Park Toronto*, 58.
40 Czerniak, 58. Emphasis added.
41 Hill, "Shifting Sites."

Bibliography

Barthes, Roland. "The Death of the Author." In *Image-Music-Text*, translated by Stephen Heath, 142–48. London: Fontana, 1977.

Berrizbeitia, Anita. "Scales of Undecidability." In *Downsview Park Toronto*, edited by Julia Czerniak, 117–25. CASE Series (Prestel Verlag). Munich; London: Prestel, 2001.

Czerniak, Julia, ed. *Downsview Park Toronto*. CASE Series (Prestel Verlag). Munich; London: Prestel, 2001.

Foucault, Michel. "Authorship: What Is an Author?" *Screen* 20, no. 1 (1979): 13–34.

Halprin, Lawrence. Lawrence Halprin on Design: RSVP Cycles, 2003. https://www.youtube.com/watch?v=QbIi966lOLs.

Hayles, Katherine. *How We Became Posthuman: Virtual Bodies in Cybernetics, Literature, and Informatics*. Chicago, IL: University of Chicago Press, 1999.

Heims, Steve J. "Introduction to 'The Human Use of Human Beings.'" In *The Human Use Of Human Beings: Cybernetics And Society*, New edition, xi–xxiii. New York: Da Capo Press, 1988.

Heylighen, Francis, and Cliff Joslyn. "Cybernetics and Second-Order Cybernetics." In *Encyclopedia of Physical Science and Technology*, 155–69. Amsterdam: Elsevier, 2003. https://doi.org/10.1016/B0–12–227410–5/00161-7.

Hill, Kristina. "Shifting Sites." In *Site Matters: Design Concepts, Histories, and Strategies*, edited by Carol Burns and Andrea Kahn, 131–55. New York: Routledge, 2015.

———. "Urban Ecologies: Biodiversity and Urban Design." In *Downsview Park Toronto*, edited by Julia Czerniak, 90–101. CASE Series (Prestel Verlag). Munich; London: Prestel, 2001.

Lettvin, J., H. Maturana, W. McCulloch, and W. Pitts. "What the Frog's Eye Tells the Frog's Brain." *Proceedings of the IRE* 47, no. 11 (November 1959): 1940–51. https://doi.org/10.1016/B0-12-227410-5/00161-7.

Lystra, Margot. "McHarg's Entropy, Halprin's Chance: Representations of Cybernetic Change in 1960s Landscape Architecture." *Studies in the History of Gardens & Designed Landscapes* 34, no. 1 (January 2, 2014): 71–84. https://doi.org/10.1080/14601176.2013.850313.

Maturana, H. R., and F. J. Varela. *Autopoiesis and Cognition: The Realization of the Living*. Boston Studies in the Philosophy and History of Science. Cham: Springer Netherlands, 1980. https://doi.org/10.1007/978-94-009-8947-4.

Ray, Thomas S. "An Approach to the Synthesis of Life." In *Artificial Life II: Santa Fe Institute Studies in the Sciences of Complexity*, edited by Christopher G. Langton, Charles Taylor, J. Doyne Farmer, and Steen Rasmussen, XI:371–408. Redwood City, CA: Addison-Wesley, 1991.

Spirn, Anne Whiston. "Constructing Nature: The Legacy of Frederick Law Olmsted." In *Uncommon Ground: Rethinking the Human Place in Nature*, edited by William Cronon, 1st ed., 91–113. New York; London: W. W. Norton & Company, 1996.

———. "Ian McHarg, Landscape Architecture, and Environmentalism: Ideas and Methods in Context." In *Environmentalism in Landscape Architecture*, edited by Michel Conan, 97–114. Washington, DC: Dumbarton Oaks, 2000.

Waldheim, Charles. "Strategies of Indeterminacy in Recent Landscape Practice." *Public*, no. 33 (2006): 81–86.

Wiener, Norbert. *The Human Use Of Human Beings: Cybernetics And Society*. New edition. New York: Da Capo Press, 1988.

Wolfe, Cary. *What Is Posthumanism?* Posthumanities Series. Minneapolis: University of Minnesota Press, 2010.

10

REFRAMING CYBERNETICS

Philosopher Yuk Hui suggests that we live more than ever in an "epoch of cybernetics" because "cybernetics was not a discipline parallel to other disciplines...but rather it aimed to be a universal discipline, able to unite all other disciplines, therefore...a universal (mode of) thinking par excellence."[1] The rise of cybernetics and its continuation as systems theory have provided a concept that appears to overcome binary thinking in many aspects: unifying nature and culture, human and nonhuman, ecology and technology, with one overarching term—system. One might use a system as a metaphor to connect entities across different scales, material or immaterial, living or non-living. In one way or another, assemblage thinking, actor-network theory, new materialism, and object-oriented ontology (OOO) all owe their reasoning to cybernetics and systems theory.

Yet, as Hui asserted, the irony is that "cybernetic thinking remains a thinking of totalization, since it aims to absorb the other into itself...which sees polarity not as oppositional but rather as a motivation towards synthesized identity," and thus "to think beyond cybernetics is to think beyond the totalizing effect of a non-dualist thinking."[2] Hui then proposes two important concepts—ecology of machines and technodiversity—which I will use to examine the cases presented in this chapter. Hui's notion of ecology exists on the basis of biodiversity; thus, an ecology of machines suggests technodiversity, in the sense that advancing one technique as the universal solution, such as using pesticides to combat insects, results in the elimination of other techniques. There was a sense of technodiversity before pesticides became the mainstream agricultural solution. Thus, as Hui argues, one way to achieve technodiversity again is through "reframing the enframing [*Gestell*]" of cybernetic thinking itself.[3] Here, Hui uses the German word *Gestell*

DOI: 10.4324/9781003320852-14

to invoke Martin Heidegger's assertion that the essence of modern technology is enframing.[4]

In the following chapters, I will explore the technodiversity of cybernetic thinking by analyzing a range of environmental practices that "misuse" technologies and "misinterpret" cybernetics. Modern control theory, as the main descendant of cybernetics, has decreased its technodiversity. Yet, I will argue that, framed within the notion of co-productive intelligence, environmental practices in contemporary art, design, and engineering provide diverse and open-ended views toward machines, thus leaving room for alternative ways to *reframe* machines and cybernetic thinking.

State-Space Representation and Feedback Control

To reframe cybernetics, we must first examine how modern control theory developed early cybernetics and formalized control systems.[5] Let dynamic system x be the function of time t, with input u and output y (Figure 10.1). Based on this diagram, controlling this system means manipulating input u to reach a desired output y. There are many ways to do that. To illustrate, I will use the example of controlling a home's interior temperature. Passive control requires no additional energy input (e.g., the walls of a building are passive controllers that stabilize interior temperatures). In contrast, active control requires energy inputs (e.g., a radiator keeps a room warm during winter). Depending on whether sensors are used, two broad categories exist within active control. Open-loop control does not require sensing system dynamics, but relies on preprogrammed control sequences to manipulate system behavior (e.g., a clock that switches the radiator on and off according to a schedule). This control method cannot account for disturbances such as exterior cold waves from the Arctic. In contrast to open-loop control, closed-loop feedback relies on a sensing system to control system dynamics. We can upgrade the radiator in the previous example by adding a temperature sensor to the room—a thermostat. Sensor data will feedback to the controller so that when the temperature drops below a pre-set threshold, the controller will switch the radiator on. The last category is feedforward control, which anticipates extraneous disturbances and feeds this information forward into the controller. In the example house, we may add a computer to monitor weather forecasts and, anticipating a cold wave, preheat the room. While these are broad categories, engineers would apply mixed methods to control a system in real cases. In my analysis, I will focus on feedback control because it is the basis of early cybernetics, and is the most common control system surrounding us, as in thermostats.

The contribution of cybernetics to modern control theory is that, by formulating a system into a black box with input and output, one need not study the internal complexities of the system in order to control it. All that

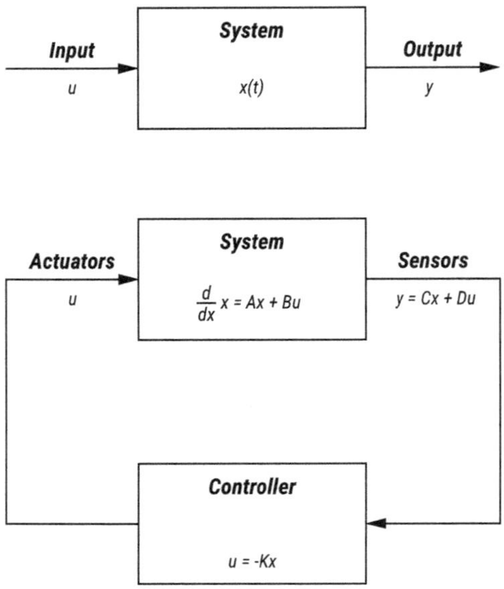

FIGURE 10.1 Feedback control. Illustration by the author.

is needed is a state-space representation of the system, and then an entire branch of mathematics is at one's disposal. A state-space representation is a mathematical model of a physical system, presented as a set of input, output, and state variables related by first-order differential equations. With state-space representation, a system could be formalized as follows:

$$\frac{d}{dt}x = Ax + Bu$$

$$y = Cx + Du$$

Here, x is a vector—an array of one-dimensional variables—representing state variables that change over time. The system matrix A describes how the state variables (x_1, x_2, x_3 ... x_n) are interrelated. u is an input vector—variables influencing system dynamics—and B is a matrix describing how the input variables u interrelate. $\frac{d}{dt}x$ is the derivative of x with respect to time t; it represents the rate of change of x. System dynamics is thus formalized as the first line of the equation; the rate of change of system variables $\frac{d}{dt}x$ at a given time is determined by state variables x and input signals u in a linear relationship. In the second line, y is an output vector that can be measured using sensors. In a perfect scenario, we assume full-state feedback in which $y = x$, in the sense that sensors can monitor full-system dynamics. As a concrete example, if the system is an airplane, A defines all the state variables related to the plane's physical design; B represents the actuators on the airplane, such as the rudder and aileron positions.

To control the system with the feedback mechanism is to design a control law $u = -Kx$ to manipulate actuators to nudge the system toward the desired state. Here, K is another matrix that defines the control law. So, u in the first function can be replaced, resulting in the following equation:

$$\frac{d}{dt}x = Ax - BKx = (A - BK)x$$

Therefore, the original system with actuators and perfect control laws may be written as a new system, in which $(A - BK)$ is a new matrix that defines system dynamics. In this equation, A is known because it represents the physical design of the system. So, mathematically speaking, by adjusting a set of control laws K associated with appropriate actuators B, we may drive the system to reach any possible state.

Controllability is thus defined in relation to the A and B matrices. In other words, if a system is not mathematically controllable, one can redesign B by adding more actuators or redesigning actuators to increase control authority. Conceptually, every column in the B matrix represents an actuator; by increasing the dimensions of B, any system is mathematically controllable. Most importantly, an extensive theory of control has been developed for linear systems supported by an entire branch of mathematics based on linear algebra. And, for certain nonlinear systems, system behavior will appear linear around a fixed, stable point. One can thus linearize the dynamics near a fixed point to stabilize a nonlinear system with a controller. Moreover, for a complex system with a high-dimensional state space (where the dimension of A is quite high), we will often observe dominant patterns closely related to system behavior. For these complex systems, control authority over these dominant structures is often sufficient to achieve an acceptable level of control.[6]

Mathematically speaking, if a system is modeled using state-space representation, one can characterize a state-space model by observing inputs and outputs and using the model to generalize the system's future patterns, based on which the system can be controlled. Modern control theory is repeatedly vindicated by engineering accomplishments and successful applications, such as aircraft, rockets, infrastructures, industrial machines, and robots. In a way, engineering successes give society the confidence to apply cybernetic thinking to further control the environment. Smart cities and cyberphysical infrastructures are proof of this confidence.

Though vastly generalized, this brief introduction demonstrates why the sensing-computing-actuating feedback loop has become the cornerstone of technological development over recent years. First, more sensors located in appropriate spots mean better estimation of environmental systems. In recent years, numerous public agencies, private organizations, and bottom-up grassroots groups have established sensing networks to gather environmental data. They hope that the resultant "big data" will reveal dominant patterns that can be computed and manipulated. Second, with improved modeling techniques, one can better define a state-space representation of the system or optimize control laws. Advanced machine-learning techniques, such as artificial neural networks, have been applied in control engineering, including in environmental management. Eventually, placing more controllable actuators in the environment will mathematically increase control authority. Notions such as cyberphysical systems encourage cities to retrofit infrastructures, such as responsive floodgates and retention ponds, with controllable pieces.

Five Premises in Control Theory

We could observe at least five underlying premises when analyzing cybernetic thinking centered around control. First, the search for stability is one of the contemporary epistemic frameworks, and control theory becomes a preeminent tool in reaching perceived equilibria. This is the legacy of modernist thinking in first-order cybernetics, where homeostasis and equilibrium have been allotted greater value. Although ideas such as emergence, chaos theory, and developmental system theory have provided ways to think about system co-productions, stability has been naturalized as a goal in controlling a system. Stability defines the notion of control.

Second, modern control theory considers intelligence as a localized capacity of an individual agent, such as a person, animal, or machine. This mirrors Norbert Wiener's 1948 title *Cybernetics: Or Control and Communication in the Animal and the Machine*. Feedback mechanisms between entities are conceived of on the basis that intelligence exists prior to any form of interaction that can be categorized as communication and control. In other words, we assume animal intelligence exists in an animal, and machine intelligence in a machine, before they can communicate with and control each other. With this conviction, environmental practices regard machines as separate from biological systems, thus perceiving machines as "artificial" while ecology is "natural."

Third, with the rise of computers, a dichotomy between the digital and the physical began to be articulated. The physical environment is where data is collected; the digital space is where computing occurs to process that data. For example, the idea of a cyberphysical infrastructure begins with the assumption that two realms exist—one digital, one physical—and that, by hybridizing them, we create coupled, arguably "smarter" systems.

Fourth, the feedback mechanism is thought to cycle through three distinct processes—sensing, computing, and actuating—partly due to the way control theory is formalized. Most modern control systems contain sensors that monitor the system, algorithms that process data and compute control strategies, and actuators that influence the system dynamics. Control theory isolates, categorizes and analyzes system operations as distinct actions. These three processes may be known by different terminologies, such as sensing-predict-control or sensing-processing-actuating, but they encompass a similar notion. *Sensing* suggests that variables are detected as digital signals represented by numerical values. *Computing* occurs when the values are fed to mathematical functions to optimize control signals. *Actuating* updates the system with controllable actuators based on optimized control laws. Eventually, these updated system states lead to a new round of sensing-computing-actuating, closing the feedback loop.

Finally, based on the feedback mechanism, learning can be defined as updating the internal models which a system uses for computing control strategies. Models represent the relationships between state variables, and need to be calibrated before being put to use. Hence, two distinct domains of expertise are established: learning and practice. Learning entails determining the underlying principles of system structures, whereas practice uses those principles to optimize control strategies. Various machine-learning techniques belong to the realm of learning; they are methods with which to calibrate or construct models based on sensed data.

Many contemporary environmental practices cannot forgo isolating the sensing-modeling-actuating feedback loop. Progress may be made on each front, to perfect tools and develop new techniques to advance the feedback mechanism itself. Over the years, sensor heads have become more accurate and compact; machine-learning techniques have grown increasingly complex; and distributed control infrastructures are more integrated. Ideas such as adaptive management promote learning by doing. Yet, that must begin with the premise that learning and doing are two entirely separate domains of operation.

Several Caveats

Let us clear several caveats before entering the cases. Although none of the examples in the following chapters illustrate a sole alternative paradigm of cybernetics, and it is doubtful that such a practice exists, these cases shed light on different aspects of an alternative framing.

Moreover, to truly appreciate the value of these cases, readers must assume a transdisciplinary mindset to take on the rather liberal understanding of a designer. We often believe that art and engineering oppose each other. Yet, this belief sidesteps the commonalities in their conceptual operations. The artists, landscape architects, and engineers discussed below have all designed cybernetic systems. From this perspective, there is little difference between an engineer and an artist because cybernetic thinking is central to both.

Most importantly, these projects involve iterative processes in which practitioners experiment, redefine questions, and seek alternative strategies. Therefore, design becomes a concept that describes the operations involved in these cases, and thus we should regard those occupied in all the cases below as designers. Using design as an umbrella term allows us to place seemingly very different cases within a transdisciplinary framework. Only in this way may we begin to map the types of cybernetic practices outside the mainstream conceptualization; they challenge the notions of agency and intelligence, denaturalize our relationship with machines and the environment, and question the role of designers and intelligent machines in the co-production of the environment.

A final caveat: although these practices are categorized based on the sensing, modeling, and actuating processes, nearly every example involves all three aspects. They are categorized as they are because they directly question a specific process. These practices also suggest alternative conceptualizations of sensing, modeling, and actuating. They designate another set of terms—coding, choreographing, and attuning—as fundamental practices in unexplored cybernetic thinking.

Notes

1 Hui, "Machine and Ecology," 58.
2 Hui, 63.
3 Hui, 65.
4 Heidegger, "The Question Concerning Technology."; Also see Chapter 3.
5 The following is a generalized discription basesed on Brunton and Kutz, *Data-Driven Science and Engineering*. Readers wish to learn more details should refer to the original work.
6 Brunton and Kutz.

Bibliography

Brunton, Steven L., and Jose Nathan Kutz. *Data-Driven Science and Engineering*. Machine Learning, Dynamical Systems, and Control. Cambridge: Cambridge University Press, 2019.

Heidegger, Martin. "The Question Concerning Technology." In *Martin Heidegger Basic Writings*, edited by David Farrell Krell, translated by Walter Lovitt, 307–41. New York: HarperCollins, 1977.

Hui, Yuk. "Machine and Ecology." *Angelaki* 25, no. 4 (July 3, 2020): 54–66. https://doi.org/10.1080/0969725X.2020.1790835.

11

SENSING AS CODING

The Episteme of the Digital Age

Sensing is the most fundamental operation in the cybernetic environment. Sensors are often considered the interface between the digital and the physical, and sensing networks are regarded as a layer of infrastructure that translates physical phenomena, such as moisture, temperature, and air movement, into electronic signals represented as ones and zeros. This data is then categorized, manipulated, and computed to inform a plethora of decisions. People tend to attribute a certain objectivity to sensing practices. It is a common belief that environmental sensing produces big data that aids us to make better decisions.

However, David Bowen's artwork challenges this conviction. Many of his installations can be described as displacements of phenomena from one place to another, like the installation "Tele-Present Wind." This installation consists of two parts. In a gallery, a series of 126 x/y tilting devices are distributed on the floor. In each tilting device, a servo motor is connected to a dried plant stalk and drives it to tilt in any direction. Outside, a "sensor" device consists of a dried plant stalk connected to an accelerometer. As the winds blow, the exterior stalk sways, and the accelerometer detects the exact intensity and direction of the movement. The movement data is then transmitted to the devices in the gallery, whose tilting stalks replicate the exact, real-time movements outside. Thus, the exterior wind is displaced into the gallery space (Figure 11.1).

Bowen's installation poses a series of questions as to the nature of sensing and sensors. Wind speed and direction are variables in meteorology, and are commonly measured via anemometers. Anemometer outputs are vectors describing wind with a combination of two values, direction and speed of

DOI: 10.4324/9781003320852-15

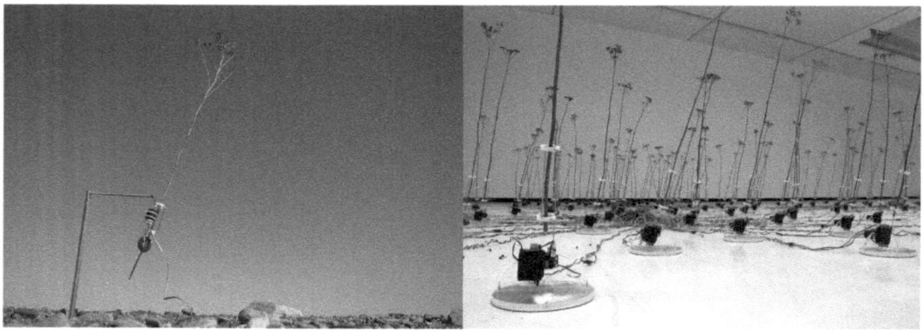

FIGURE 11.1 David Bowen, "Tele-Present Wind" (2018); sensor (left) and actuators (right). Courtesy of David Bowen.

horizontal airflow. However, what is sensed about wind in Bowen's installation? Alternatively, what constitutes a sensor? Bowen's sensor is an accelerometer. Its output is a vector combining two values—direction, and magnitude in that direction. Bowen used a two-axis accelerometer in order to record how fast and in which direction the wind tilted the stalk. Therefore, technically, the accelerometer did not sense wind directly, but recorded the stalk's movement in terms of direction and acceleration.

Compare Bowen's sensing device with the three-cup anemometer commonly used in environmental sensing. There is, in fact, no difference between them in how a phenomenon is translated. A cup anemometer consists of three hemispheric cups mounted on horizontal arms. When air passes the cups in any horizontal direction, the cups drive a vertical shaft to turn at a rate proportional to the wind's speed. Thus, counting the turns over a set time interval yields a value proportionate to the average wind speed during that time frame. Hence, a cup anemometer itself does not directly measure wind speed; what it actually measures is how fast its shaft rotates. In both cases (a cup anemometer and Bowen's accelerometer on a stalk), devices can produce certain behaviors in response to wind—the rotation of a shaft or the tilting of a stalk. We interpret the phenomenon by reading the devices' behaviors.

Bowen's artwork exemplifies that any type of sensing involves interpretation. Here, I will use Don Ihde's hermeneutic relationship between humans and technology to further illustrate (see Chapter 3). Sensors as technological artifacts compress and black-box a set of interpretation protocols into a single, streamlined process. In hermeneutic relationships, we are involved with the environment via an artifact that provides a representation of the world. We must interpret its behavior in order to gather information about the environment.[1] Ihde uses hermeneutic relationships to describe this type of mediation process, because the artifact must be "read." Yet in order to read, a set of protocols must be fashioned to accompany the device's specific design.

For example, anemometer factor, or the ratio of wind speed and shaft rotation speed, depends on the dimensions of the cups and arms. Further, sampling frequency indicates how often an anemometer takes measurements: a 1 Hz frequency anemometer counts the turns once per second, and a 0.1 Hz frequency counts the turns made in 10-second intervals. Interpretation thus involves specific designs of the devices and protocols that instruct one to read the device "properly." Construction of a sensing device also means standardizing and formalizing a wide range of protocols, procedures, and possible designs, with one working model.

Thus, a sensing device does not innocently "listen" to a phenomenon; it is, in essence, a black box containing streamlined interpretation procedures. Charting the history of cybernetics and the emergence of the "digital," media historian Bernhard Siegert formulated the relationship between mathematics and cultural techniques. "The mathematical concept of the symbol is founded on the black-boxing of a history of cultural techniques, which in turn is a history of the articulation of the real."[2] A sensor is an idealized way to read environmental traces. Thus, sensing implies coding the environment, by naturalizing and streamlining a history of interpretation techniques with an automated recording device.

We approximate an observed phenomenon with a sensing instrument's behaviors, assuming that studying the quality of the measuring device is equivalent to studying the original phenomenon. In the case of an anemometer, we replace air movement with shaft rotation. The behaviors of sensing instruments thus become proxies of the phenomena in question. This replacement of objects applies to any form of environmental sensing. One of the most common temperature sensors is the thermistor, or thermal resistor. Its resistance changes drastically with temperature fluctuation; thus, a temperature sensor constructs a relationship between voltage and ambient temperature. Similar principles apply to light sensors or photoresistors whose resistance changes with light levels.

Therefore, sensing is coding, which naturalizes and legitimizes a series of interpretation protocols of translation that encode the environment into ones and zeros. From this vantage, environmental sensing constructs a network of various instruments, each of which embodies a specific history of interpretation techniques, coding the environment into a "datascape." When we interact with the environment using environmental data, we are, in fact, interacting with this "datascape."

Sensing and Its Objectivity

How is this "datascape" weighted with a sense of objectivity, and why would we trust the encoding process to be a faithful depiction of environmental phenomena? Modern environmental sensing continues historians Lorraine Daston's and Peter Galison's notion of the paradigm of *mechanical objectivity*. Tracing the development of nineteenth-century scientific atlases and representations, Daston and Galison posit that the first phase of objectivity, or "truth-to-nature," relates to efforts to capture nature "as it was" through meticulously detailed representations of biological, geological, and other phenomena, to render them both recognizable and classifiable.[3] However, the evident problem was the artist-authors' need to tame nature's variability into idealized forms and archetypes. With the popularization of daguerreotypes arose the era of *mechanical objectivity*.

> By mechanical objectivity we mean the insistent drive to repress the willful intervention of the artist-author, and to put in its stead a set of procedures that would, as it were, move nature to the page through a strict protocol, if not automatically. This sometimes meant using an actual machine....[4]

On many levels, environmental sensing may be understood as "a relentless search to replace individual volition and discretion in depiction by the invariable routines of mechanical reproduction."[5] On the path for "truth-to-nature," scientists became their own enemy, and "the all-too-human scientists must, as a matter of duty, restrain themselves from imposing their hopes, expectations, generalizations, aesthetics, even ordinary language on the image of nature. Where human self-discipline flagged, the machine would take over."[6] Mechanical objectivity is thus a paradigm whereby scientists develop machines and routines of mechanical reproduction that remove human interpretation from scientific research and "let nature speak for itself." It is not only an issue of accuracy but also one of scientific morality and epistemic virtue, and machines are simultaneously means and symbols of this paradigm. "The machine stood for authenticity: it was at once observer and artist, free from the inner temptation to theorize, anthropomorphize, beautify, or interpret nature."[7] Sensors as automatic recording devices become paragons for the virtues of ideal observers; they are "patient, indefatigable, ever-alert, probing beyond the limits of the human senses."[8] Sensors provide data "free" from human interpretation; they remove human observers from the equation, and provide a perspective that issues from no one and nowhere.

When we place the development of a sensing instrument within sociotechnical networks, we find that the sensing practice is a self-vindicating process. Eric Winsberg argues that computer simulation practices in scientific research are self-vindicating over time, and thus simulations possess a life of their own.

> As simulation practices evolve and are retooled, the techniques they employ carry with them their own history of prior successes and accomplishments, and when properly used, they can bring to the table independent warrant for belief in the models they are used to build.[9]

Just as computer simulations gain their reliability by carrying with them an entire history of accomplishments, sensing practices may also be understood as an evolutionary process through which credentials are gained by repeating successful methods and assumptions. A sensor, too, carries its own history of scientific and engineering accomplishments.

From a social constructionist perspective, the evolution of a sensor involves multiple social groups, over a prolonged period, agreeing on one standardized version and a set of accepted sensing protocols. Each successful prediction based on the produced data adds to the reliability of the sensing device's underlying principles and protocols. Because data produced by anemometers are repeatedly proven to be useful in predicting weather patterns, the underlying assumptions and protocols of anemometers were, over time, legitimized. When standards and protocols are set in place and shared by stakeholder communities, including environmental groups, manufacturers, distributors, and hobbyists, environmental data is warranted not only by individual humans but also by a socio-technical ensemble too sprawling to trace. Again, objectivity is gained by creating a view from no one and nowhere.

Mediation and Limitation: The Datascape and the Episteme of the Digital Age

Sensors allow us to relate to the environment through mediation, which simultaneously limits us to interactions in the specific ways permitted by the sensors, accepting the sensors' history of cultural techniques. We build sensors to code the environment into sensible variables and describe the environment as an object. Rather than sensing wind, we code airflow movement with instruments, and construct an object of the wind to communicate this phenomenon with each other, using comparable qualities. In this process, we reduce a phenomenon to a limited number of variables that fit within meaningful and quantifiable categories, such as wind speed and direction, rather than other aspects for which we do not yet possess vocabularies. Through the black-boxing of a history of cultural techniques, sensors become the physical and material manifestation of a set of relationships between us and the environment.

Thus, by analyzing how a sensor works, we may analyze how we are coupled with the environment through the relations permitted and simultaneously limited by the sensor. In developing a measuring instrument, we simultaneously eliminate other possibilities with which to construct relationships with processes exterior to ourselves. When a sensing instrument becomes standardized, we begin to ignore the fact that each sensor embodies a history of interpretation technique; we overlook the assumptions inserted into building such a device and fix our eyes on the output it produces.

For example, an anemometer measures only horizontal airflow, but winds may be three-dimensional; a horizontal wind profile thus becomes a proxy for wind in general. Further, streamlining protocols and the self-vindicating process eliminate other possibilities of using or misusing sensing instruments to differently code the environment. The earliest mechanical anemometer dates to the fifteenth century, and it has engendered numerous versions. Today, many anemometers use mechanisms other than rotating cups to measure wind speed, such as ultrasonic, hot-wire, laser Doppler, and wind pressure. If we were to transform Bowen's device into an anemometer, we might use mathematical functions to translate accelerometer data into wind speed and direction. To a certain degree, as long as a contraption responds to wind with salient behavior, its outputs can be coded and translated into wind speed and direction. The movements of willow twigs can be coded to measure wind. However, the complexity of their movement is beyond human capability to record, categorize, quantify, and compare. We might say we invented anemometers to measure wind speed. Yet we can also say that we failed to develop vocabularies to describe the movement of willow twigs, because it is easier to build a machine to categorize such a phenomenon, and it requires less human interpretation on the path to objectivity. However, by doing so, our language for discussing wind is also confined to direction and speed as the two recognizable and classifiable variables allowed by anemometers. These assumptions associated with sensing practices as specific cultural techniques have been black-boxed into the sensing devices. They become a hidden epistemic substrate that one must accept without question.

By embracing environmental sensing, one also consents to the blind spots associated with sensors. As Siegert declares, coding is "the most basic cultural technique of the digital age, in the sense that it is based on an isolation of the nonsensical and its declaration as nonreality..." and when coding occurs, "the nonsensical becomes the nonexistent."[10] To become "digital," one must accept that the "datascape" constitutes one's whole reality, and declare whatever exists in the blind spot of a sensor as nonexistent. Scholar Ranjodh Singh Dhaliwal argues that "addressability" is the foundational requirement for computation.[11] In order to be computable, a phenomenon must be addressable, in this case, by sensors. What is "unaddressable" becomes invisible to sensing devices, yet it still exists. For example, when a self-driving car collides with a pedestrian, the datascape that constitutes the car's reality renders the pedestrian invisible, thus utterly incomputable. From the car's perspective, there are no pedestrians, because they have become unaddressable by the sensors.

Addressability differs from quantifiability. One may construct protocols and proxies to quantify any phenomenon, making it addressable, and then computable. Unaddressability, however, involves the situation in which phenomena are ignored as noise or categorized as uncertainties; they exist in the blind spot of the epistemic framework that underpins sensing practice.

To attribute objectivity to sensing, we also must trust whatever data is produced by sensing as the preeminent way in which the world is revealed to us. Sensing becomes coding, a principled way of knowing. If coding is the most basic cultural technique of the digital age, then undergirding our time—the one characterized by digitalization, big data, and machine intelligence—is an episteme built around coding, in which "nonsensical" become "nonexistent" and "nonreality."[12] This is an episteme of the digital age, because sensing is no longer a practice employed solely to construct scientific evidence. Sensing practices have permeated every aspect of contemporary society, and formed a sort of technological momentum, a concept in social construction of technology (SCOT). An emerging technological artifact or system develops in a social network and eventually couples with it, resulting in the seamless mesh of a socio-technical ensemble. Increasingly, sensing practice becomes an attractant that structures our relationships with other species and the environment, forming a sort of technological momentum that forces us to couple with the environment in a manner that is both enabled and limited by sensors.

Environmental sensing proponents' enthusiasm for this enterprise cannot be separated from an inherent liberating narrative. Safecast[13] exemplifies the paradigm of environmental sensing, and has been placed on many labels over the years, such as citizen science, open-source, grassroots, and bottom-up. After the 2011 Fukushima nuclear disaster, public and governmental agencies were unable to provide the public with data regarding radiation levels. To address this issue, hackers and technologists formed a team to develop DIY radiation-sensing kits. The team designed the kits, contacted manufacturers, and organized volunteers to collect and upload data to a web-mapping platform, producing timely visualizations of the radiation levels around Fukushima. Over the years, Safecast has evolved from its original focus on radiation levels to broader environmental concerns, such as air quality.

In 2020, Safecast developed a web map to track COVID-19 cases, hoping to provide transparency regarding critical information often held by government agencies, and to promote a data and knowledge democracy. Further, in July 2022, Safecast accounted its "bGeigies4Ukraine" initiative to bring people's attention to Russian troops' activities at Ukrainian nuclear facilities, which created fear and concern in Ukraine and neighboring countries. Environmental sensing is therefore intertwined with DIY culture, citizen science, and liberating actions that challenge authorial control over knowledge production. Simultaneously, participating in sensing practice becomes a way to demonstrate concern as a responsible global citizen, and gives a sense of agency so that accumulated individual actions will culminate in meaningful changes. Environmental sensing narratives fit well within the contemporary culture and an ideology of individualistic democracy.

This phenomenon can be described as a combination of smart mentality and environmentality.[14] Environmental sensing and smart environment narratives make sensing practice a disciplinary strategy. To become a concerned and responsible citizen who cares about the environment, one must participate in "smart" activities, such as environmental data collection. A growing technological momentum expressed as environmental values intertwined with democratic ideologies might mobilize different social sectors to participate in constructing a global sensing network. However, this global network will simultaneously construct a "datascape" that replaces the environment in which we live and limits the possibility of constructing alternative relationships with the environment, by imposing a specific epistemic framework in which the nonsensical is nonexistent. Consider the notions of co-production and co-evolution. Humans have co-evolved with tools and other technical systems, co-producing each other. Yet on the reverse side of co-evolution and co-production between humans and sensing technologies is a process of devolution; we have eliminated other potential pathways through which the environment is revealed to us, and denied other possible ways in which we might be coupled with nonhuman species. If we understand sensors as media enabling us to interact with the environment, then, at the same time, sensors limit the ways in which we can connect with the environment. In regard to environmental sensing, embarrassingly, mediation and limitation share the same process. This is not a critique of environmental sensing practices, and we should recognize that examples such as Safecast have generated positive impacts. In addition, it seems the discursive liberating narratives play a more important role than the produced datasets themselves. As we are obsessed with the accumulation of big data and its potential for liberation, should we not reflect on the fact that it limits other potential pathways to liberation by restricting the ways in which we could interact with the environment?

Here, we must recognize the limitations in employing critical and historical frameworks of analysis. Historians, including Daston, Galison, and Siegert, have provided meticulous analytical frameworks to reveal a shocking aspect of naturalized notions, such as sensing; yet, in the end, dissatisfied readers ask, "What comes next, and can we do something about it?" Moving forward, we must take on a design research framework to recognize the opportunities embedded in sensing practices.[15] The construction of a "datascape" brings terror as well as hope, because there remain so many different ways to code the environment other than what is framed by the mainstream sensing practices that focus on control and communication.

Coding Operations as Co-production Techniques in the Digital Age

When reframing sensing as coding with which one can actively construct relationships with other objects or assemblages, we might intentionally code the environment into datascapes that are liberating and inspiring. Thus, sensing as coding becomes a co-production technique specific to the digital age. "Co-production is shorthand for the proposition that the ways in which we know and represent the world (both nature and society) are inseparable from the ways in which we choose to live within it."[16] In other words, to frame sensing as coding entails an intentional decision to use or "misuse" sensing instruments in specific ways so that the environment may be revealed differently. Such knowledge can transform our lived relationships with others. In a way, coding is about reframing modern technology's "enframing (*Gestell*)" effect as articulated by Heidegger. Thus, how do we code the environment in such a way as to cultivate greater empathy toward the nonhuman realm and foster co-productive relationships with other species in a shared environment?

Developing Dhaliwal's notion of addressability and unaddressability, I propose a third category—misaddressing—to expand the spaces of operation in sensing practice. In order to be computable within a cybernetic environment, a phenomenon must first be visible in the datascape. One way to make something visible in a digital age is to create difference by reconfiguring pathways in the system, thus misaddressing, thereby producing errors and noise. We can visualize misaddressing through a series of diagrams (Figure 11.2). As shown in the first diagram, if sensors are interfaces between the digital and the physical, sensing networks become membranes between the environment and its datascape. Further, I will employ a body of artwork to challenge this diagram, with four categories of misaddressing practice: cloning (one-to-many), assembling (many-to-one), rerouting (one-to-one), and physicalization (one-to-naught).

Cloning (One-to-Many)

Berlin-based artist Simon Weckert's performance art "Google Maps Hacks"[17] best illustrates the one-to-many relationship. Google Maps uses GPS pings and the accelerometer in a smartphone to determine its user's location and speed. Users carrying smartphones thus become temporary sensors distributed in an urban environment. Google Maps uses these user-phone sensors to optimize navigation instructions. In early 2020, Weckert carried 99 smartphones in a handcart while he strolled the streets of Berlin. This action deceived Google Maps into believing that traffic jams had congested the streets. Wherever Weckert walked, he created virtual traffic jams, turning green streets red on Google Maps. These virtual traffic jams might have created physical impacts if the algorithm rerouted drivers around the indicated streets, in which only one man pulled a small handcart with 99 smartphones (Figure 11.3).

Here, the relationship between Weckert with his 99 smartphones and Google Maps is neither addressable nor unaddressable. A Media Access

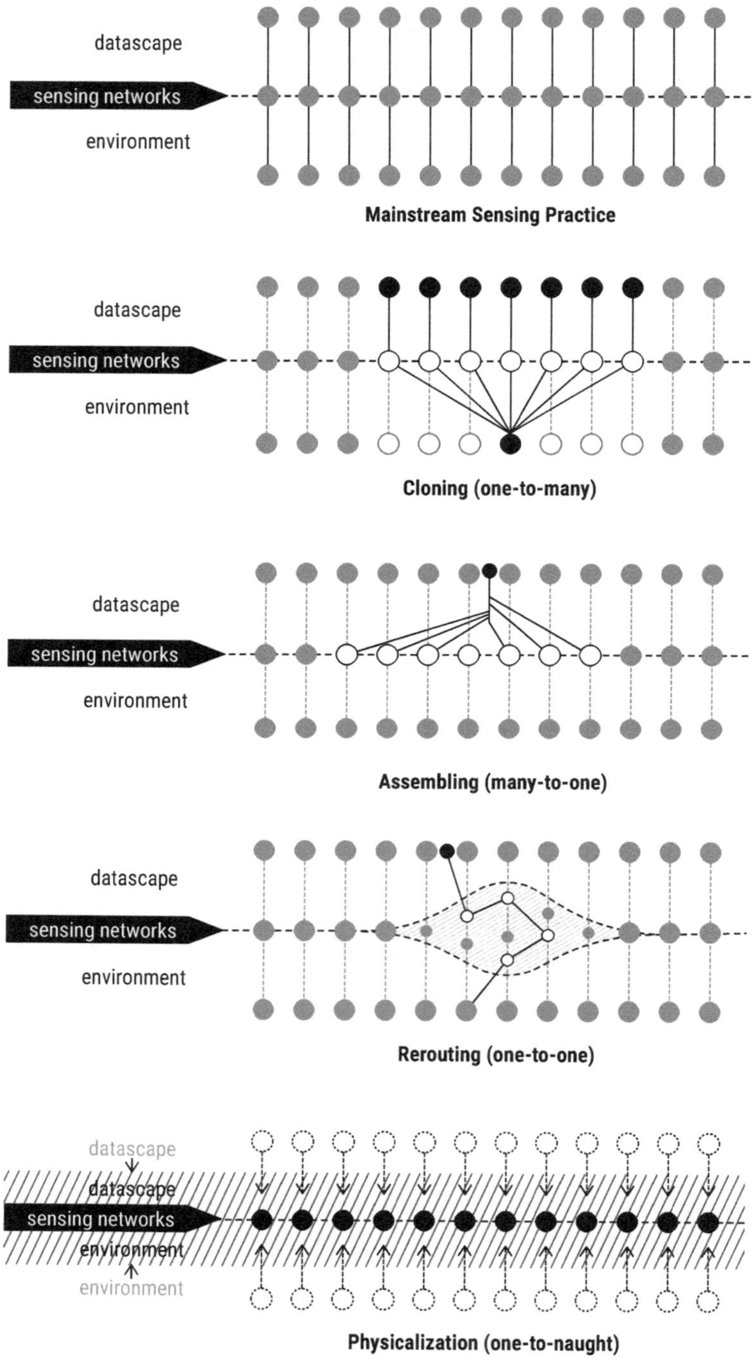

FIGURE 11.2 Misaddressing. Illustration by the author.

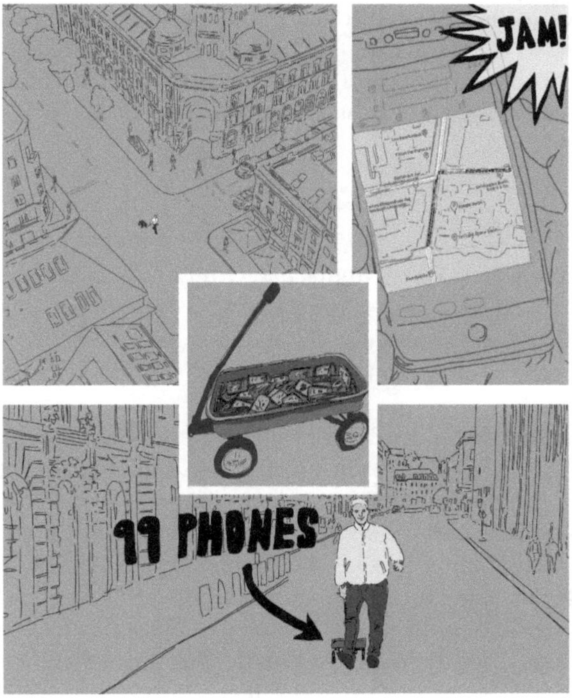

FIGURE 11.3 Simon Weckert, "Google Maps Hacks" (2020). Illustration by George McCracken.

Control (MAC) address is a unique identifier for a device on a network. Google Maps recognized the smartphones' MAC addresses and located them via GPS signals. Thus, these smartphones were addressable. Yet there is another level of addressability. Google Maps tracks users through smartphones as media, and each user is replaced by a MAC address. Smartphones become the proxies of users, so Google Maps presumed a one-to-one relationship between a user and a smartphone. Weckert, however, assigned one user with 99 MAC addresses in 99 phones, creating a one-to-many relationship. Misaddressing created digital duplicates, overloading Google Maps to produce "errors."

Google's team responded to this art performance with good humor, thanking the artist for his creative use of Google Maps and promising to improve their algorithms. Certainly, the true value of the project lies beyond helping Google Maps improve its algorithms. The artist actively codes the datascape that constitutes Google Maps' full reality. By doing so, the artist opens the black box of the Google Maps sensing network, denaturalizing the invisible protocols and assumptions. In a cybernetic environment, all phenomena, including human behaviors, are transformed and reduced by sensing devices into computable variables. Algorithms modulate, manipulate, and compute these variables in order to optimize the physical environment and human behaviors. Yet Weckert's art suggests that we may create digital clones in a cybernetic environment to reveal its invisible algorithms, protocols, and assumptions. Denaturalizing a cybernetic environment is the first step toward reframing it.

Assembling (Many-to-One)

Sougwen Chung's artwork "Omnia per Omnia" denaturalizes surveillance practices and reimagines landscape painting as a collaboration between an artist, a robotic swarm, and a city's dynamic flows. Rather than a view from a single perspective, her "landscape painting" captures a multidimensional view of the city.[18]

Quoting AI researcher Fei-Fei Li, who stated that, "If we want our machines to think, we need to teach them to see," Chung taught her drawing robots to "see" the world through computer vision algorithms. In collaboration with Nokia Bell Labs, Chung and her team trained a computer vision algorithm—called Motion Engine—with internet videos from publicly available camera feeds, using a technique called optical flow to characterize and quantify the motion of objects in a video scene. With this technique, the artist and her team analyzed the collective density, direction, dwelling, and velocity of urban movement—pedestrians and vehicles flowing through urban spaces. These variables were extracted as positional data that became paths for her robots to draw on. In a way, these robotic paths represent not the movement of any individual, but the movement of a swarm.

> The philosophical underpinnings of the Bell Labs Motion Engine captures the optical flow of a scene as opposed to the single object; it privileges the action of the collective (the behaviour of the crowd) over individual surveillance (face-tracking and recognition). The latter way of seeing [is] fragmented and discrete.[19]

Over the past years, many data visualization projects entertained the notion of "flow of a city" as a proxy for dynamic urban patterns, such as "One Day on Waze" and "Foursquare Check-ins."[20] In those cases, each data point on the maps represented a Waze or Foursquare user. Their geolocation data was collected by the GPS unit in their smartphones. There was a one-to-one relationship between each person and each dot, while at the same time, those who switched off location services, or did not use the apps, became invisible in the "flow" maps. When "Foursquare Check-ins" claimed the maps showed "the pulse of New York City," one comment below the video reminds the creators that the visualization recorded only "the pulse of Foursquare users."[21] This comment embodies the ongoing debate concerning the tensions between participatory sensing and individual data privacy, since "crowd-sourced" data does not at all reflect the dynamics of the "crowd."

In contrast to the one-to-one visualizations of "flows," Chung created a many-to-one relationship that challenges the notion of addressability. Instead of addressing each user as a data point, many-to-one is a conundrum in which we are not sure to whom we should pin an address. An interesting phenomenon emerges: no one is invisible since the crowd is being watched and analyzed by machines, yet meanwhile, everyone is invisible—because individuals become less important within each movement of a swarm. Furthermore, swarms are ephemeral; if the crowd falls into complete chaos, the swarm disappears and the computational address pins to nothing. Moreover, the concept of individual agency fails in a swarm since we are unsure how much a single agent's movement influences the swarm's behavior. As Chung stated in an interview:

> The movements [of the robots] are linked to the movements of a crowd, who aren't aware of their position as catalysts of the robotic swarm. But what if the crowd were aware of being all watched over by these machines? That's what I mean by being at the onset of this collective imagination and intersubjectivity. It's inspired the idea of a collaborative space between multiple bodies that is physical, digital, telepresent, and most importantly, shared. *It's a representation that isn't about control, but something else…* The project really stemmed from a curiosity—a willingness to de-privilege the Western conception of the individual towards an entangled, intersubjective, radical ecosystem. I've been inspired by this reconfiguration of the "I", through theorists like Yuk Hui who explore a new cosmotechnics, and the philosophies of media centred around eastern concepts of relation.[22]

Yuk Hui defines cosmotechnics "as the unification of the moral and the cosmic through technical activities," and it "reopens the question of moral beyond ethical rules which are added posteriorly as constraints to new technologies."[23] Chung's "Omnia per Omnia" challenges us to completely rethink the current ethical and moral debates on sensing technologies. It teases out the epistemic framework of sensing that conceives *individual* as the fundamental unit for its operation, as well as for its ethical and moral critiques. "Omnia per Omnia" bypasses discourses based on data privacy, state surveillance, and individual freedom, all inevitable reasonings stemmed from an individualistic view toward sensing and computation practices. Instead, it reframes sensing and computation on the basis of crowds and collectives and provides novel ways to discuss phenomena in societies such as China, where data tracking, facial recognition, and surveillance systems seem to vex individuals less than in Western societies, and are occasionally welcomed. Consider how distinctly China and the United States used tracking apps to handle the early stage of the COVID-19 pandemic. The discourse surrounding data privacy and state surveillance, which relies on a conception of a duality between the individual and the state apparatus, often falls short when addressing cases like COVID tracking. This is not to compare methods, but to illuminate that current discourse oversimplifies the Chinese case based on a Western conception of individuality. As Chung noted, the surveillance practice in her art is no longer about control, but something else, which we only begin to conceptually explore through practices like her artworks.

Rerouting (One-to-One)

Another case illustrates the rerouting relationships that construct new pathways in a cybernetic environment. Mileece is a sonic artist who makes music with plants. In her installation "iOracle," she attached electrodes to the leafy limbs of plants to capture micro-bioelectricity from the leaves.[24] Electrodes are biomedical sensors for measuring human heart, muscle, and brain activities, such as in EEGs and ECGs. In "iOracle," Mileece connected the electrodes to bodies of plants rather than humans. The micro-current passed through an amplifier and was transformed into binary codes to animate melodies and harmonic frequencies utilizing custom music software. When plant leaves were touched by audience member or ants, direct feedback in the form of music was produced. Mileece hoped that the sonification of the bioelectric signals could cultivate empathy in her audiences, so they would acknowledge different forms of sentience.

Mileece thus rerouted data transmission pathways and built a cascade of unusual connections between sensors, electronic parts, and algorithms. The relationships between the datascape and the objects were still one-to-one, while a new pathway was constructed by associating one address with another in ways that appear abnormal, even outrageous. The artist produced ones and zeros from the plants as datasets that failed to fit within existing data categories. Her artistic practice exposed the blind spot in the "datascape" constructed by present-day mainstream sensing practices, and elucidated aspects that are unaddressable yet existent.

Bowen's installation mentioned above, "Tele-Present Wind" also falls within this category. By attaching accelerometers to a dry plant stalk, Bowen produced a dataset that rested outside the computable variables of wind.

In both cases, artists have denaturalized a "sensor" as the interface connecting physical space to digital space. Instead, both reimagine sensors as assemblages constructed of materially different components—whether mechanical or biological. In Bowen's case, an accelerometer with a dry plant stalk becomes a "wind sensor" that elucidates aspects of wind beyond speed and direction. Similarly, for Mileece, electrodes plus plant leaves become sensors to detect human touch and crawling ants, constructing a new feedback loop between species.

When we search for touch sensors at off-the-shelf sensor distributors, we find two types. One type is the capacitive touch sensor, based on capacitive coupling, which can detect anything that is conductive or has a dielectric difference from the air, such as smartphone touchscreens. This type produces a binary presence/absence relationship, whether touching or not. Another type of touch sensor is based on the force-sensitive resistor (FSR). These FSRs are resistors that alter their resistive value depending on the surface force applied. Compared to both sensor types, Mileece's "touch sensor" made of plant material produces data about neither case. For the audience, touching the plant's leaves became conversations with the plant-sensor itself.

Rerouting practices treat the space between digital and physical as an area of intervention.

In these rerouting practices, a sensing network is no longer a thin membrane between the digital and the physical where a phenomenon is digitized into ones and zeros. Instead, the artists unbox the moment of sensing and elongate the data transmission process, reconfiguring the interpretation protocols black-boxed by sensors.

Physicalization (One-to-Naught)

Dietmar Offenhuber's artwork and theoretical reflections attack the accepted dualism between the digital and the physical in sensing practices. He proposed *autographic visualization* as a speculative counter-model for conventional data visualization methods that begin with data collection and end with charts explaining the underlying patterns in datasets.[25]

"*Staubmarke*" (dustmark) is a public artwork in Stuttgart, Germany, a city much affected by particulate matter pollution.[26] Contrasted with conventional workflow that addresses environmental issues by visualizing collected data on a web platform, "*Staubmarke*" constructed physical evidence of air pollution. Using a technique called reverse graffiti, the installation crew created images of dotted patterns by strategically removing dirt on urban surfaces, such as walls and bridges, with a high-pressure water jet. Over time, these graffiti marks will fade as particulate matter re-accumulates on their surfaces (Figure 11.4).

Another project, "Ozone Tattoo," used indicator plant species as "sensors" to call attention to climate change and its impacts. Climate change creates conditions that increase ground-level ozone, which is harmful to plants, animals, and humans. Low-ozone concentrations damage tobacco plants, "tattooing" yellow and brown spots on their leaves, which makes the tobacco plant an apt indicator species with which to visualize the effects of climate change. This installation established ozone gardens, which allowed communities to monitor pollution by observing the indicator plants (Figure 11.4).

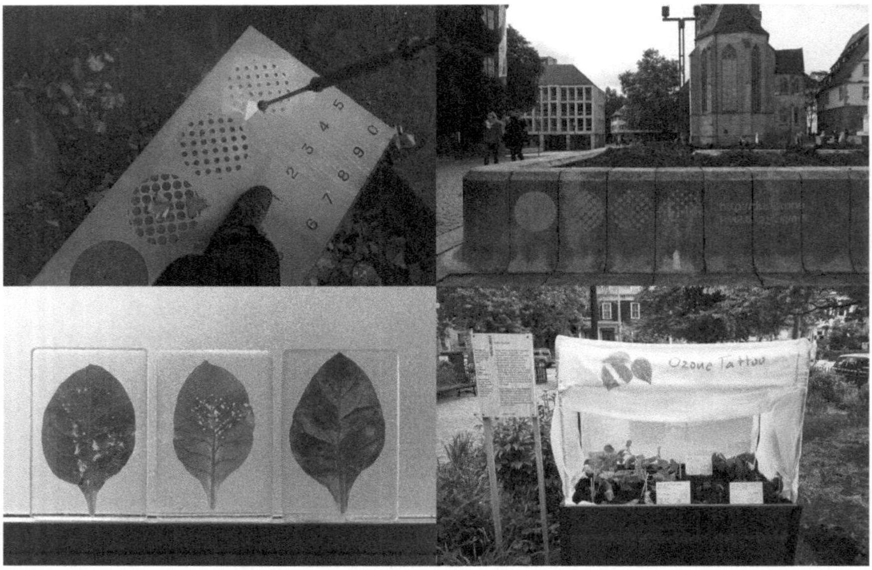

FIGURE 11.4 Dietmar Offenhuber, “Staubmarke” (2018) and “Ozone Tattoo” (2019). Courtesy of Dietmar Offenhuber.

Recall Daston and Galison’s discussion on *mechanical objectivity*. Data collection uses sensing devices to construct scientific evidence allegedly free from human interpretation; the nature of the phenomenon may thus be presented “as it was.” However, as argued, sensing *is* coding, and a sensor does not innocently “listen” to these phenomena. Instead, changes in the environment trigger a set of system operations within a sensor—a series of interpretation protocols—to produce certain output, which we call data. In order to make sense of the phenomenon, one must read the traces constructed by the sensor; this occurs through hermeneutic relationships allowed by the sensors. The traces make sense only when interpreted in certain semiotic relationships. One must construct a mental model for interpretation in order to understand that 50 ppm of PM2.5 signifies a relatively acceptable air quality. In contrast, the “reverse graffiti” and “Ozone Tattoo” created systems that present the phenomenon directly to us. There was no data visualization, since there no “data” was digitized. Both projects created a one-to-naught relationship within the misaddressing framework, because nothing was produced in the so-called “datascape.” In both cases, environmental sensing produced physical traces in the environment. The digital and the physical collapsed into each other, and the environment became data.

Coding, Design, and Climate Change

Reframing sensing as coding shifts our inquiries regarding issues such as climate change. With sensing, questions are limited to what types of sensors and how to distribute them in the environment to collect more data to create better predictive models. With coding, however, the question becomes how to code the changing climate into signals that can register differences within present-day sociopolitical systems. Based on Gregory Bateson's definition that "information consists of differences that make a difference,"[27] the present-day sensing regime fails to produce any "information." How might "climate change" be a politically sensitive and debatable issue, when scientists have presented numerous datasets suggesting accurate facts? This dilemma is less a failure of political persuasion than a failure to code climate change to register differences in contemporary sociopolitical systems. Therefore, environmental sensing regarding climate change must concern itself not only with collecting more data, but also with coding to construct information that can be registered by political, cultural, and societal systems. Only in this way will they begin to react to this pressing issue.

Timothy Morton conceptualizes climate change as a hyperobject that exceeds human perception in space and time, and is thus inaccessible.[28] Meanwhile, Harman proposes that metaphors, which construct unusual relations between objects, become valuable pathways through which we connect with other objects.[29] From an OOO's perspective, coding practices highlight environmental sensing as a preeminent tool in developing new metaphors between objects. The art installations detailed above shed light on strategies to construct metaphors through sensing practices, and connect us with the hyperobject climate change. For example, "Ozone Tattoo" connects climate change to ozone-induced plant disease, and, from there, to community and citizen science.

However, to recognize the transformative values of these art practices requires us to overcome the notion that art is the direct opposite of science, and thus they are incommensurable paradigms. A common thread in the above art projects involves misusing tools to produce "freaks" that are "useless" in mainstream environmental sensing practices, and may be categorized only as "artistic expressions." "Creative use of sensing technology" is merely a euphemism for "useless" within the mainstream sensing paradigm, which emphasizes objectivity, efficiency, prediction, and control. Ironically, the notions of originality and idiosyncrasy, commonly treated as virtues in the production of art, become obstacles that hinder these practices from being truly transformative. The concept of "artistic practices" overemphasizes individual creativity and original thinking, and overlooks underlying similarities among different artists in terms of their conceptual strategies. From this perspective, between mainstream sensing practices in science and various counter-practices in art, designers could assume an important role in reducing this disciplinary gap.

The four types of coding operations based on the notion of misaddressing—cloning, assembling, rerouting, and physicalizing—serve as strategies for designers to replicate these art projects beyond mimicry. They attempt to translate these artworks into conceptual strategies with which to intervene in the cybernetic environment on a systemic scale. Clearly, these four types of coding operations are not intended to encompass all sorts of counter-practices, but my goal is to provide examples for an analytical design research framework that transforms knowledge from one realm to another. This book is thus a continuation of landscape architects' decades-long tradition of transforming the strategies of art practices into public landscapes. Landscape theorist Elizabeth Mayer argues that landscape architects have successfully translated environmental values into landscape practices, via environmental arts, in the decades since the first Earth Day in 1970.[30] Today, when a cybernetic environment becomes the new condition for environmental practices, cases from digital art serve as valuable lessons for the next generations of landscape designers.

Finally, we must reflect on the notion of intelligence of co-production, technodiversity, and the ecology of machines. Yuk Hui regards biodiversity as the basis for ecology, and thus the ecology of machines recognizes technodiversity in machines and techniques. When sensing practices are awash with claims of producing objective big data to predict and control the environment, we simultaneously eliminate other techniques to relate to the environment and other species beyond prediction and control. Environmental sensing, in the name of science and objectivity, has black-boxed a set of cultural techniques in specific ways. Mainstream practices prevent the possibility of other cultural "sensing" techniques thriving. Continuing the metaphor of ecology in sensing practice, the cases presented in this section produce "mutations" in the contemporary monoculture of environmental sensing. They appear to be abnormal instances within the environmental sensing ecosystem. Yet mutations are essential for evolution, because mutation provides a pool of random genes for the ecosystem to grow diverse. Furthermore, intelligence emerges from interactions between assemblages and objects, through relationships such as loose coupling, adversarialness, and symbiosis. Within the monoculture of sensing practice, machine intelligence is defined by techniques of optimization and efficiency, eliminating other forms of machine intelligence. Thus, the ecology of machines also denotes a diverse framework for machine intelligence. Once mutations are introduced, they become catalysts for the emergence of other forms of machine intelligence, beyond optimization and control.

Notes

1 Ihde, *Technology and the Lifeworld: From Garden to Earth.*
2 Siegert, "Coding as Cultural Technique," 10.
3 "The Image of Objectivity"; *Objectivity.*
4 Daston and Galison, *Objectivity*, 121.
5 Daston and Galison, "The Image of Objectivity," 98.
6 Daston and Galison, 81.
7 Daston and Galison, *Objectivity*, 139.
8 Daston and Galison, 139.
9 Winsberg, *Science in the Age of Computer Simulation*, 45.
10 Siegert, "Coding as Cultural Technique," 18.
11 Dhaliwal, "On Addressability, or What Even Is Computation?"
12 Siegert, "Coding as Cultural Technique."
13 https://safecast.org/.
14 Vanolo, "Smartmentality"; Gabrys, "Programming Environments"; Krivý, "Towards a Critique of Cybernetic Urbanism."
15 Designers and scholars may hold different interpretations of "design research." My perspective is rooted in the conviction that it constitutes a transdisciplinary practice, capable of mobilizing ideas from one field to another.
16 Jasanoff, "The Idiom of Co-Production," 2.
17 Weckert, "Google Maps Hacks."
18 Chung, *Omnia per Omnia.* https://sougwen.com/project/omniaperomnia. Last accessed: April 29, 2024.
19 Chung.
20 Waze, *Data Visualization - 8 Cities. Foursquare, Foursquare Check-Ins Show the Pulse of Cities.*
21 https://vimeo.com/75413842. Last accessed: April 29, 2024.
22 Chung, *Interview with Ravelin Magazine.* Emphasis added.
23 Hui, "Machine and Ecology," 64.
24 Mileece, *iOracle.* https://www.mileece.is/ioracle. Last accessed: April 29, 2024.
25 Offenhuber, "Data by Proxy -- Material Traces as Autographic Visualizations."
26 Offenhuber, *Staubmarke (Dustmark).* https://offenhuber.net/project/staubmarke-dustmark/. Last accessed: April 29, 2024.
27 Bateson, *Mind and Nature: A Necessary Unity*, 99.
28 Morton, *Hyperobjects.*
29 Harman, *Object-Oriented Ontology.*
30 Meyer "The Post–Earth Day Conundrum: Translating Environmental Values into Landscape Design."

Bibliography

Bateson, Gregory. *Mind and Nature: A Necessary Unity*, 1st ed. New York: Dutton, 1979.

Chung, Sougwen. Interview with Ravelin Magazine. Interview by Jillian Billard, July 5, 2018. https://sougwen.com/interview-with-ravelin-magazine.

———. *Omnia per Omnia.* 2018. Mixed media. https://sougwen.com/project/omniaperomnia.

Daston, Lorraine, and Peter Galison. *Objectivity.* New York; Cambridge, MA: Zone Books, Distributed by the MIT Press, 2007.

———. "The Image of Objectivity." *Representations* 40 (October 1, 1992): 81–128. https://doi.org/10.2307/2928741.

Dhaliwal, Ranjodh Singh. "On Addressability, or What Even Is Computation?" *Critical Inquiry* 49, no. 1 (September 2022): 1–27. https://doi.org/10.1086/721167.
Foursquare. *Foursquare Check-Ins Show the Pulse of Cities*. 2013. https://foursquare.com/infographics/pulse.
Gabrys, Jennifer. "Programming Environments: Environmentality and Citizen Sensing in the Smart City." *Environment and Planning D: Society and Space* 32, no. 1 (February 2014): 30–48. https://doi.org/10.1068/d16812.
Harman, Graham. *Object-Oriented Ontology: A New Theory of Everything*. London: Penguin, 2018.
Hui, Yuk. "Machine and Ecology." *Angelaki* 25, no. 4 (July 3, 2020): 54–66. https://doi.org/10.1080/0969725X.2020.1790835.
Ihde, Don. *Technology and the Lifeworld: From Garden to Earth*. Midland Book. Bloomington: Indiana University Press, 1990.
Jasanoff, Sheila. "The Idiom of Co-Production." In *States of Knowledge: The Co-Production of Science and the Social Order*, edited by Jasanoff Sheila, 1–12, 2004.
Krivý, Maroš. "Towards a Critique of Cybernetic Urbanism: The Smart City and the Society of Control." *Planning Theory* 17, no. 1 (February 1, 2018): 8–30. https://doi.org/10.1177/1473095216645631.
Meyer, Elizabeth K. "The Post–Earth Day Conundrum: Translating Environmental Values into Landscape Design." In *Environmentalism in Landscape Architecture*, edited by Michel Conan, 22:187–244. Washington, DC.: Dumbarton Oaks Research Library and Collection, 2000.
Mileece. *iOracle*. 2018. Mixed media. https://www.mileece.is.
Morton, Timothy. *Hyperobjects: Philosophy and Ecology after the End of the World*, 1st ed. Minneapolis: Univ Of Minnesota Press, 2013.
Offenhuber, Dietmar. "Data by Proxy -- Material Traces as Autographic Visualizations." *arXiv:1907.05454 [Cs, Math]*, July 11, 2019. http://arxiv.org/abs/1907.05454.
———. *Staubmarke (Dustmark)*. October 5, 2018. Mixed media. https://offenhuber.net/project/staubmarke-dustmark/.
Siegert, Bernhard. "Coding as Cultural Technique: On the Emergence of the Digital from Writing AC." *Grey Room* 70 (March 2018): 6–23. https://doi.org/10.1162/GREY_a_00236.
Vanolo, Alberto. "Smartmentality: The Smart City as Disciplinary Strategy." *Urban Studies* 51, no. 5 (April 1, 2014): 883–98. https://doi.org/10.1177/0042098013494427.
Waze. *Data Visualization - 8 Cities: One Day on Waze | Waze*. October 2, 2014. https://www.youtube.com/watch?v=WMzftwQuawM.
Weckert, Simon. "Google Maps Hacks," 2020. http://www.simonweckert.com/googlemapshacks.html.
Winsberg, Eric. *Science in the Age of Computer Simulation*. Chicago, IL: University of Chicago Press, 2010.

12

THE RISE OF INTELLIGENT AGENTS

A Non-Model-Centric Paradigm

This chapter will follow a specific model—the Storm Water Management Model (SWMM)—and observe how SWMM's role in managing urban environments has evolved when new modeling techniques were introduced: intelligent agents emerged as important actors in the cybernetic environment. I present the materials in this chapter based on published papers, interviews with experts, and my long-term engagement with a scientific research group at the University of Virginia, including participation in its weekly meetings, engagement in discussions, and collaboration on research papers.

Between 1969 and 1971, SWMM was developed in response to society's growing concerns regarding water pollution. Since then, it has been widely used as a decision-making tool in planning, managing, and designing urban drainage systems. A consortium of contractors, including Metcalf & Eddy, Inc., the University of Florida, and Water Resources Engineers, Inc., developed the first version of SWMM, under the sponsorship of the Environmental Protection Agency (EPA). The goal was to create a comprehensive mathematical model capable of representing urban hydrological processes, such as stormwater runoff and combined sewer overflow, to assist administrators and engineers in planning, evaluating, and managing urban hydrology. Users may employ the model to simulate stormwater system performance from inlet to outlet, with options to exchange intermediate water storage and treatment devices. The model also provided economic data for cost-benefit analyses of alternative management policies and infrastructures.[1] In the decades following its introduction, SWMM enjoyed several major updates and became increasingly complex. The latest version (SWMM 5), released in 2014, added a climate change add-on and low-impact development (LID) functions, such as green roof, bioretention, swales, and permeable paving.[2]

DOI: 10.4324/9781003320852-16

Civil engineers and planners use SWMM to simulate the urban runoff scenarios of new developments under different rain events, and to design stormwater infrastructure schemes. They may further simulate schemes with different variables, such as culvert size and routing options. Control units, such as retention ponds and detention ponds, can be added and simulated to evaluate their efficacy in mitigating flooding. In addition, SWMM can be used to test LID infrastructures, such as filter strips, bioretention ponds, and infiltration trenches, to encourage groundwater recharge and mitigate urban water pollution.

Systems thinking undergirds SWMM, simultaneously limiting how the urban environment is conceptualized, visualized, and eventually constructed. Data moving through SWMM are rendered into hydrographs and pollutographs. A hydrograph represents the rate of water flow in a time series at a node in a system. Similarly, a pollutograph displays the concentration of pollutants as a function of time at a given node. Let us take computing hydraulics as an example. In SWMM, different components are conceptualized as sub-systems, with their own input and output. Each component accepts a hydrograph as input and processes it, producing another hydrograph as output. These individual components are models, with variables describing different characteristics of the infrastructures or spatial features. For example, a storage unit may be described by properties like maximum depth, surface area, and evaporation factors. These properties become variables involved in computing output hydrographs based on input data. For instance, to describe a storm event, a rain gauge component is used to produce a hydrograph as the input of the whole system via a catchment component. A catchment component uses the output from the rain gauge as input, passing data through a function—with variables such as slope, imperviousness, roughness coefficient, and infiltration—to describe how water collects in the catchment's watershed. A second hydrograph is produced as output. A link component, such as a culvert, connects a catchment to a node component, such as a retention pond. Regulator components, including orifices, weirs, and pumps, may be added between link and node components to control water flow speed from one to the other.

Clearly, through systems thinking, one can abstract and divide a complex urban drainage system into discrete, manageable, and solvable equations. One must construct multiple hierarchies, by imagining input and output points in a continuous process. First, one must draw a conceptual boundary, often defined by the boundary of a development area on a continuous landscape and define a point of the outlet where water eventually flows outside this boundary. Then, within the study area, a second hierarchy is introduced to separate a continuous water pathway into links and nodes. With these hierarchies in place, one isolates a group of phenomena from a continuous process and makes them analytical components, such as culverts or retention ponds.

One can further model each component, using mathematical functions, study them in isolation, and parametrize them through testing and verification. After the components are individually modeled, they may be connected. In fact, in the technical report for SWMM Version 1, scientists described their modeling process in a similar fashion by describing how each portion of a drainage system was individually modeled.

However, we must admit that this is not the only way a drainage system can be modeled and represented. In theory, one could construct a 3D digital model of the physical space and run particle-based simulations by applying physics to each water droplet to observe how the droplets interacted with the digital environment. This technique is called computational fluid dynamics (CFD) and is widely used in visual effects and in the gaming industries to simulate water, liquids, fogs, and clouds. Another option is to build a scaled physical model and pour water onto it. Before computation technologies were widely adopted, the Army Corps of Engineers constructed massive physical models to determine flood risks. On the outskirts of Jackson, Mississippi, lies a 200-acre physical model of the entire Mississippi River Basin, from Pennsylvania to Montana and south as far as Louisiana. Constructed in the 1940s, and in operation from 1949 to 1973, the model was used by engineers to test rainfall events and make reinforcement decisions. Engineers also simulated the performance of proposed infrastructures and ran tests to assist in developing computer models.[3] However, after SWMM's development, it quickly dominated within the ecosystem of models for urban hydrological management because it fits better within a descriptive and quantitative epistemic framework in modern science and engineering.

The mediation and limitation effects of machines are again manifest in modeling. The use of SWMM makes it possible to comprehend and interact with highly complex phenomena, through the mediation effect. Yet, specific choices made when creating SWMM also define how urban drainage systems are studied. There are at least two aspects to consider.

First, on a systems level, there is no distinction drawn between a retention pond and a wetland reservoir, in terms of water storage. When they are conceptualized as storage units, their volume is described with the same model parameter, called, in both cases, "storage curve." This curve abstracts the shape of a storage unit by describing how the surface area changes with water depth. Then, SWMM uses the curve to compute the stored volume as a function of depth. In other words, from SWMM's perspective, wetland areas, retention ponds, and underground storage tanks are each represented as plots of storage curves, despite their material differences, as well as their different ecosystems and the lives they sustain. This representation renders elements in the conceived system both replaceable and dispensable.

Second, from a design perspective, it may be easier to understand a continuous process as discrete procedures, but the environment consists of a continuous field of different ecosystems. Through systems thinking, one is limited to designing a system as if building with Lego blocks, connecting pieces based on input and output without worrying about engaging the site as a whole. Analyzing from a territorial scale, the entire site is itself a storage unit, with varying degrees of dryness and wetness. To a certain extent, model-making processes reinforce the notion that a storage unit must be a physical object within the environment, thus compelling us to build one that meets the parameters based on simulated results. Because a storage unit is represented in SWMM as a nodal component, both computationally and graphically, one assumes it to be an object with a defined shape and boundary. Yet, puddles formed in street potholes and depressions might be viewed as distributed storage units across the roadway. Unfortunately, they are incomputable in SWMM because they do not fit within the quantifiable and descriptive framework allowed by its model-making practices. Consequently, these ephemeral puddles fail to become conceptual forms in the vocabularies for designing and constructing the environment, despite the possibility of utilizing puddles to store water and anticipating emergent street ecologies.[4] In the epistemic framework allowed by SWMM's specific method of model-making, things outside the descriptive frames and parameters are categorized as uncertainties to be controlled.

The limitation of model-making is not about abstraction or oversimplification, but constraining thinking with one universal epistemic framework, which conceptualizes the environment as a system of manipulable components with parameters and rules for connections. When we embark on a specific route of model-making, we simultaneously reduce technodiversity, giving up on other ways to model the environment, such as physical simulation or CFD models, and even on a non-model-centric way to think about the environment.

Consider Eric Winsberg's discussion regarding scientific computer simulation, in which he argues that simulation practices are "self-vindicating" and "have their own lives."[5] As Winsberg noted,

> ... [simulation practices] evolve and mature over the course of a long period of use, and they are 'retooled' as new applications depend more and more reliable and precision techniques and algorithms...Whenever these techniques are employed successfully—that is, whenever they produce results that fit well into the web of our previously accepted data, our observations, the results of our paper-and-pencil analyses, and our physical intuitions; whenever they make successful predictions or produce engineering accomplishments—their credibility as reliable techniques or reasonable assumptions grows.[6]

Here, "a life of its own" should not be confused with technological determinism, asserting that technologies possess their own trajectories of development independent of society. Instead, Winsberg specifically emphasizes that "lives" refers to "the whole host of activities, practices, and assumptions that go into carrying out a simulation."[7] What Winsberg articulates is a co-production and co-evolution process between humans and machines. When building SWMM, selecting one specific parameter in the modeling process might have involved art and intuition, but these intuitive choices were verified by their success in applications. Every successful prediction and engineering achievement created with SWMM will eventually become part of its history of credibility. This specific parameter will then carry into SWMM's new versions and updates, and will thus be naturalized as a "gene" in this specific model species. Co-production manifests in the sense that this specific "gene" makes SWMM a necessary and effective medium through which engineers engage with the environment. Of course, necessity and effectiveness are evaluated under certain criteria, characterized by efficiency and productivity, that embody an epistemic framework of modern society.

Over the past years, SWMM has become increasingly accessible. The first generation of SWMM was written in Fortran, a programming language developed by IBM in the 1950s specifically for scientific and engineering computing. However, SWMM5 is essentially a rewrite with C, a widely used general-purpose programming language that allows the model's wider integration into other fields. Because SWMM5 is open-source, its source code has been integrated into different software, such as AutoCAD Civil, as a backend simulation engine. Moreover, PySWMM provides a Python interface to the SWMM5 model. Python is a widely used programming language within machine-learning communities, and its popularity generates intersections between environmental engineering and artificial intelligence research.

We will follow a group of scientists from the Link Lab at the University of Virginia (UVA) to analyze how interdisciplinary inquiries transform the role of modeling in environmental practices. In January 2018, UVA's School of Engineering established the Link Lab as a research incubator to connect its different disciplines, such as computer science, electrical and computer engineering, and systems and environmental engineering, to collaborate on research in cyberphysical systems. One of the many projects at the Link Lab was the Data-driven Management for Interdependent Stormwater and Transportation Systems (dMIST). Its team was composed of instructors and Ph.D. students from civil engineering, computer science, and transportation engineering, plus government officials from Norfolk, Virginia. By that time, I was a Ph.D. candidate at UVA's School of Architecture, working on my dissertation concerning cybernetics and environmental design. As a doctoral researcher, I was invited by a Science, Technology and Society (STS) scholar on the dMIST team to participate in their weekly meetings, conduct joint research projects, and, most importantly, "follow the engineers"—using Bruno Latour's term—to study how engineering disciplines conceptualize a cybernetic environment.

The dMIST team sought to use cybernetic technologies to couple the stormwater and transportation systems of coastal cities, such as Norfolk, to create an urban environment more resilient to climate change and sea-level rise.[8] With an overall sensing-predicting-controlling framework, the dMIST team's approach to environmental management was based on cybernetic thinking and real-time control. Environmental data collected through sensing networks and crowd-sourced datasets could be applied to construct machine-learning models, which could then be used to predict future scenarios and support decision-making.

Before continuing, let us examine dMIST's background information and underlying premises. Most urban stormwater infrastructures are passively controlled; culverts, retention ponds, and weirs exist passively to respond to flood events. Certain cities have built controllable floodgates that can be closed during severe floods. Due to climate change and rising sea levels, many coastal cities now experience more nuisance flooding, the temporary inundation of low-lying areas during high-tide events, especially under full and new moons. Sea-level rise also accompanies increased tidal fluctuation; certain outfalls may be submerged during high tides, resulting in regular backflow into the stormwater system. In addition, climate change carries unexpected weather patterns and severe storm events that threaten cities' flood infrastructures.

With these increased environmental uncertainties, coastal cities such as Norfolk search for alternative ways to be adaptive and resilient, including incorporating smart technologies. These efforts include updating stormwater systems with real-time control strategies by installing sensing networks and retrofitting stormwater infrastructures with controllable actuators. For example, StormSense is a project initiated by scientists from the Virginia Institute of Marine Science (VIMS), collaborating with cities in the coastal Hampton Roads metropolitan area, including Norfolk. The StormSense team has installed dozens of sensors across these cities in order to monitor water levels.[9] Cities, including Norfolk, have also contracted with emerging companies such as Opti, which specializes in environmental monitoring and real-time control, to update their stormwater systems and management platforms. Thus, dMIST is based on the premise that society's enthusiasm for smart city technologies, and pressing environmental crises such as climate change, will eventually propel future cities to adopt more cyberphysical infrastructures. In other words, a new type of cybernetic infrastructure will emerge in coastal cities. The environment will be laden with machine intelligence waiting to be mobilized by engineers and designers. In a way, dMIST's research agenda embodies a broader cybernetic imagination of society. The team searches for alternative ways to engage with the urban environment with cyberphysical infrastructure, through a cybernetic framework articulated as sensing-predicting-controlling feedback loops.

The dMIST team possesses multiple sub-teams with projects focusing on different aspects of the research, and I will use two projects as examples. These projects may be understood as a seriation that uses machine-learning algorithms to determine real-time control policies for cyberphysical stormwater infrastructures. The first project uses model predictive control (MPC) as a feedback control framework; the second project utilizes a reinforcement learning (RL) algorithm. Both strategies are tested on a simulated stormwater system built with SWMM5, and the system is inspired by a real-world scenario located in the City of Norfolk (Figure 12.1). We will refer to this as a "toy system."

This simple toy system consists of two sub-catchments, representing two different urban areas connected to a pair of retention ponds. The outlet is a body of water within a tidal zone. Junction 1 (J1) is considered a storm drain located on a roadway. In addition, J1 may flood if a high tide causes tidal backflow and prevents water from discharging. The overall goal is to prevent flooding in both retention ponds and J1.

In the passive-control scenario, at the outlet of each pond is a weir at a fixed elevation so that the water level in each retention pond is fixed and cannot be emptied. In the real-time control scenario, however, these two weirs are retrofitted with controllable valves at the bottom of the pond side. In this case, if weather forecast data indicates a looming storm event, and real-time sensing data shows the tide is low enough, then operators—a human or a machine-learning agent—can open the valves to discharge water,

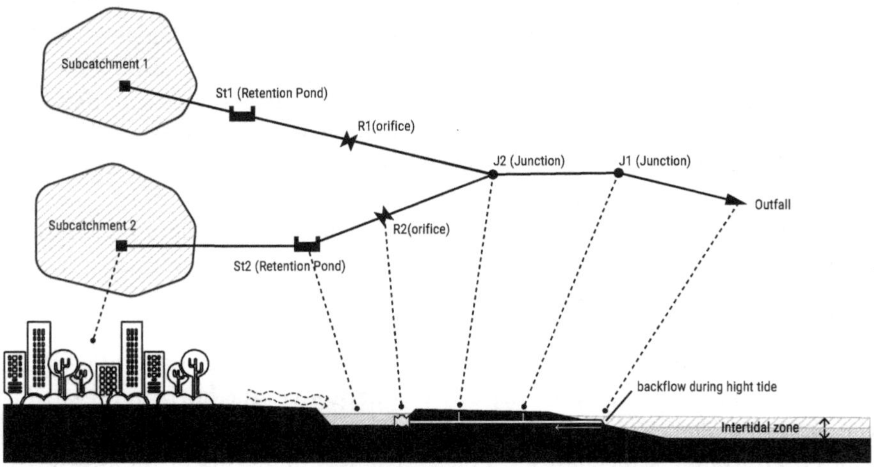

FIGURE 12.1 Simulated stormwater system in SWMM for MPC and RL experiments. Illustration by the author based on Bowes et al., "Flood Mitigation in Coastal Urban Catchments Using Real-Time Stormwater Infrastructure Control and Reinforcement Learning."

lowering the level in the retention ponds to make room for incoming floods. Contemporary best practices for real-time control are carried out through rule-based control, which can be visualized as a decision tree consisting of a series of if-then statements based on experience and calculations from experts and human operators. For example, if a forecast presages a storm event that might cause 1,000 m^3 flooding, then the valve must be opened 100% to drain an equivalent amount of water, plus a 20% safety factor from the retention pond. Both MPC and RL are ways to determine control rules using machine-learning algorithms.

There is one caveat before we continue. In both the MPC and RL cases, a model built with SWMM5 (Figure 12.1) was used as a toy system to replace the physical environment for experimental purposes. This establishes the same environmental baseline for comparing the MPC and RL cases because the only difference is the controller in the simulated watershed.

Model Predictive Control (MPC)

Controlling a system may be understood as reducing errors—the gaps between the desired system state and the current system state. Effective control strategies gradually move the system to the desired state. To optimize control strategies, MPC has become a cornerstone of modern process control. It emerged in the 1970s and was first used in process industries in the 1980s; it is now widely used to control all sorts of feedback systems.

In the initial project carried out by the dMIST team, its scientists implemented MPC in managing the toy system in Figure 12.1.[10] As the name suggests, MPC, or model predictive control, requires a predictive model to simulate and predict outcomes of control strategies in order to pick the best one to execute in the real system. In the dMIST example, the scientists conveniently duplicated the SWMM5 toy system as the predictive model. With this predictive model, the MPC algorithm simulates different actuation strategies forward in time and optimizes these strategies based on objective functions. At each step, the MPC algorithm looks ahead for a brief time window, or horizon, to determine the immediate control strategy. Computation costs limit the intervals between each time step for control strategies. This project uses a 15-minute time step—this means the actuators (the positions of two valves) can be adjusted every 15 minutes.[11] Once this control action is implemented within the system, the MPC controller moves the horizon forward to the next time step to repeat simulation and optimization. Essentially, MPC relies on a feedback mechanism: in each round, the results of the previously implemented control actions are accounted for in computing the next immediate control action. Yet, MPC differs from simple feedback controls for which the control action is based solely on past errors. When deciding its next move, MPC instead looks both behind and ahead of time. For example, a thermostat is a simple feedback machine that evaluates only the gap between the current and the desired temperatures to decide whether to switch air-conditioning on or off. If we implemented MPC on a thermostat, we might build a thermodynamic computer model of the room to simulate and optimize a range of control strategies (e.g., the output temperature and intensity of the system's fan) in a given horizon before implementing the optimal control strategy. Using MPC here is, of course, overkill in controlling room temperature, since it is computationally expensive. The dMIST experiment was conducted on a high-performance computer with a 20-core CPU. Further, MPC is considered an "online" controller, meaning the system is constantly connected, monitored, and simulated by the MPC algorithm in real time.

For the dMIST team, the goal was to prevent flooding in the system, and, at the same time, maintain target water levels in both retention ponds. A genetic algorithm was introduced as an optimizer to search for the best control strategy in each time step. Genetic algorithms are akin to natural selection, in which the best qualities of a generation are passed to the next generation. In each generation, an array of policies is deployed, and the best are retained. In the next generation, mutations are introduced to successful policies, and a new array of updated policies is deployed. Repeating this selection-mutation process optimizes the algorithms' performance. In this project, at each time step, the MPC algorithm casts an array of ten different control strategies as the first generation and runs ten parallel SWMM simulations to find the best policies. These ten policies in the first generation are then optimized with the genetic algorithm by repeating the selection-mutation process for another five generations. Finally, the optimal strategy is implemented in the toy system. This will update the system and initiate the next round of optimization. In the dMIST example, MPC combined with a genetic algorithm simulates not only multiple scenarios but also multiple generations of simulations. This significantly increases computational cost but improves control outcomes.

Therefore, MPC is model-centric because the key to success is whether the model used to simulate and optimize strategies best captures the physical environment's characteristics—regardless of whether it is a physics-based model such as SWMM, or a physical exemplar such as the Mississippi Basin model. However, a model merely estimates, no matter how complex or accurate it is, and never becomes the modeled system. Even in the case of a large physical simulation, such as the Mississippi Basin model, a scaling factor is introduced as a variable for abstraction, since everything but water shrinks in the model. In the above MPC example, the physical environment is replaced by a model created with SWMM5, which is simultaneously used as the predictive model for finding strategies. This presents a non-existing perfect scenario, where the model is a perfect replica of the environment, and no gap exists between the two (or full-state feedback in which $y=x$ in the state-space representation in Figure 10.1). However, if this MPC-based strategy were put to use in a real cyberphysical system, a supercomputer running SWMM would always be connected to a real stormwater system to simulate different scenarios and devise strategies. Thus, SWMM becomes a "digital twin," a digital replica of the physical system. Ideally, the control policy would be more effective if the digital twin used in the simulation better captured reality. Here, the deciding factor for a successful MPC is whether a perfect digital twin can be constructed. Thus, MPC is model-centric, since it challenges us to produce better models to approximate the environment, reducing a gap that can never be closed.

Reinforcement Learning (RL)

Driven by the rise of big data, high-performance computing, and advanced machine-learning algorithms, such as artificial neural networks, data-driven modeling and control are undergoing a revolution.[12] In cases such as MPC, machine-learning techniques can directly learn a state-space model of the system using only data, thus bypassing physics-based models. Unlike a physics-based empirical model, such as SWMM, state-space models may be linearized to characterize the system dynamics. Because linear functions are faster to compute than non-linear functions, latency between the system and the controller is greatly reduced, and computing speed is thus increased. Furthermore, MPC requires running full simulations in parallel; it is therefore essential that the model runs at accelerated speeds. The Link Lab scientists noted that their MPC strategy could benefit from learning a state-scape model, to replace SWMM and more quickly find control strategies.[13] Yet no matter how fast a model runs, MPC is a model-centric approach that requires a model, no matter its form and complexity, to compute control strategies.

However, the emergence of big data and the widespread adoption of advanced machine-learning algorithms have given rise to a non-model-centric paradigm. As discussed in Chapter 8, recent developments in AI research have introduced transformative cases that call for reevaluating our understanding of intelligence. Notably, RL agents such as AlphaGo and AlphaStar have not only outperformed human players but have also developed strategies with a distinct "machine flavor." Training an RL agent is to iterate the process of trial and error; the agent will first carry out a random action and will be rewarded if the action can effectively achieve goals defined by the computer scientist. Because machine learning is considered a black box, RL agents can occasionally produce unexpected policies.

In the second project carried out by the dMIST team, an RL agent was implemented to control the toy system by altering the positions of two valves connected to the retention ponds.[14] A popular algorithm for RL is called deep Q-network. Here, "Q" stands for quality, or how useful a given action is in gaining rewards at a given state, while considering future actions in gaining future rewards. A "q-value" is thus represented as an action-state function *Q(s, a)*, with *s* for state and *a* for action. A "q-table" organizes many q-values into a matrix, with rows for states and columns for corresponding actions. Thus, training a q-learning agent means finding the optimal q-table. After training, the agent may thus be understood as an optimized q-table, with a reward value for each corresponding action in each state. Deep Q-networks are therefore restricted to a finite number of discrete actions (e.g., a game controller possesses limited buttons for the agent to push). In the case of valve control, the action space for this agent is continuous and infinite; the valve can be altered to any position, from 100% closed to 100% open. In valve control, a deep Q-network is unsuited to the task.

The dMIST experiment employed a different RL algorithm known as Deep Deterministic Policy Gradients (DDPG), specifically designed to address problems with infinite input spaces, such as valve positions.[15] The DDPG agent comprises two key components: an actor and a critic. The actor assesses the current environmental state and determines an appropriate action, while the critic scrutinizes the pairing of the environmental state and actions, ultimately assigning an action score. To initiate a training session, the SWMM provides state information to the actor, including data such as pond water levels, tide levels, and forecasted rainfall. Subsequently, the actor formulates an action that modifies the SWMM model. This action can either ameliorate the situation by averting floods and maintaining desired levels in retention ponds or exacerbate it by causing flooding or deviating water levels from their intended targets. The pairing of the actor's action and updated state information is then passed on to DDPG's critic, which assesses the action's effectiveness. Based on the critic's evaluation, the actor updates its action in the next round to maximize rewards.[16]

The dMIST scientists used real-world rainfall and tidal data to produce a range of flood events that trained and tested the RL agent. The agent learned policies for adjusting the two valves to prevent upstream flooding by directing water into the retention ponds while avoiding the release of water during high tide, which would cause flooding downstream. The agent also learned to maintain specific target water depths in the two retention ponds. In this research, no human knowledge regarding the stormwater system was given to the RL agent. To some extent, the RL agent developed its own understanding of the system.

The scientists compared this RL agent with other control strategies, including passive control (no control), rule-based control, and the above-mentioned MPC control. The RL agent not only outperformed human rule-based control and MPC but also executed interesting policies that inspired designers to rethink urban stormwater management. One apparent example occurred when the RL agent would open or close the valves multiple times over a short period to discharge water during a rain event, causing multiple insignificant flood events downstream in J1. This strategy does not exist in current rule-based control, and at first glance is counterintuitive, since the best way to control flooding is to retain as much water as possible in retention ponds.

To understand why the agent devised this strategy, we need to analyze the reward function used in training the agent. Recall the example of the pancake-flipping robot in Chapter 8. Experimenters designed an objective function to reward the agent for each second the pancake stayed in the air. However, the robotic arm would throw the pancake as high as possible to maximize the immediate reward. Objective functions do not inevitably lead to expected behaviors. We can observe similar behavior in the dMIST's RL example. The objective function was designed as follows:

$$r = \begin{cases} -\Sigma(\text{flooding}), & F > \delta \\ -J1_{\text{flooding}} - (|St1_{\text{depth}} - \text{target}| + |St2_{\text{depth}} - \text{target}|), & F = 0 \end{cases}$$

In the function, δ represents a forecast rainfall threshold, which divides the objective function into two parts indicated by $F > \delta$ and $F=0$. In the first line, whenever forecasted rainfall in the agent's observable horizon exceeds the threshold ($F > \delta$), the agent would gain a higher score for reducing flooding throughout the entire system. In the second line, if there is no forecasted rainfall ($F=0$), the agent is rewarded for maintaining the target water depth in the retention ponds, and for not flooding the downstream node J1 by doing so.[17]

Upon closer examination, such a reward function carries an inherent bias because it assesses flooding based solely on volume, overlooking frequency. The RL agent is not punished for the frequency with which it causes flood events, but only for their overall volume. This leaves room for the RL agent to game the function. The agent can gain reward by maintaining the target level in the retention ponds, even though this requires discharging water from them and causing several low-volume flood events downstream at the J1 node. Doing so, the agent would incur negative rewards. However, these low-volume events result in negligible negative rewards compared to the positive rewards gained from maintaining water levels in the retention ponds. The RL agent exploited this bias in the function to devise an unconventional strategy.

This does not mean we must implement another function to produce an unbiased agent. Instead, the situation asks us to reflect on the ways in which stormwater systems have always been optimized based on limited parameters. The agent's behavior in exploiting the frequency of flood events to achieve its goals reminds designers that we overlook an entire school of strategies, based on frequency rather than volume, especially when considering cyberphysical systems that allow for live updates and real-time responses.

Of course, the above function is not the sole route to articulating goals. In fact, an early version of the experiment used a different function that rewarded the agent if it maximized the water levels of the retention ponds during rain events. There, reward was not directly associated with flooding events, and minimizing flooding was inherent in maximizing the ponds' water levels. The scientists tested multiple ways to articulate their goals through reward functions. This process was similar to communicating with a human system operator through formalized functions and observing how the operator performed with the instructions. As we see in the RL example, something different happened; this is no longer about control, but something else. To frame this difference, we need to consider the possibility of a model-free paradigm.

From MPC to RL, with the rise of data-driven approaches and machine-learning techniques using artificial neural networks, we observe the emergence of intelligent agents. In conceptualizing the role of modeling in control, they operate in two different paradigms. In modern control theory, any control problem may be generalized with a standard feedback system. The current system state is monitored by sensors and fed back to a controller, which then uses control laws to compute actuation signals, driving the system toward the desired state. Thus, control relies on first modeling the system to determine a function that best represents the system dynamics. In other words, modern control theory is based on the premise that to control a system, one must determine a model of it.

However, the RL example above questions this premise by operating in a model-free framework. Thus, RL begins by conceptualizing an intelligent agent made of deep neural networks capable of making decisions based on observed information, and adjusting its strategies based on continuous observation. The RL strategy bypasses system modeling and does not rely on the perfection of a digital twin. Instead, the RL experiment begins with the controller as an agent, which tries to learn a control policy by interacting with the system and observing how the system responds to these interactions.

This is oddly similar to human learning, such as when we learn to drive automobiles. We do not begin by studying a car's internal physics and mechanics; instead, we press the pedals and turn the steering wheel, trying to acclimatize ourselves to driving. Here, "acclimatizing" entails directly interacting with the car to observe its general patterns and how it responds to our driving strategies. Because learning to drive begins with the agent rather than a model, the result is not optimized control laws, but a diverse range of nuanced personalities and driving strategies. Consider how different drivers stop at a red light—some gradually slow, while others slam their brakes. Just as pilots and astronauts learn to operate their crafts on simulators, SWMM becomes a virtual gymnasium where the agent may develop all sorts of strategies. The results reflect different personalities and nuances. The RL agent in the example above was cautious and proactive in managing the retention ponds. Whenever it saw looming storm events, it discharged water in the retention ponds to prevent flooding, even though this was unnecessary and caused minor downstream flooding.

Training a machine agent is akin to coaching basketball players. To achieve their goals, coaches must articulate different offensive and defensive strategies to their players. Nevertheless, the players are those who interpret and carry out strategies, and the coach must relinquish a sense of control; the coach operates from a higher order. Engineers are involved in a similar relationship with an RL agent. Before training the agent, the engineers must first design the agent's deep neural network architecture. These artificial neural networks are designed as layers of connected neurons. Thanks to Python-based machine-learning libraries, such as TensorFlow, OpenAI Gym, and Keras, designing a deep neural network has become extremely simple. One needs only several lines of code to call certain types of layers from the library and connect them into a fully functional artificial neural network. Ben Bowes, a Ph.D. researcher, built the RL agent in the above experiment. When I asked him how and why the RL agent was constructed in this experiment, he noted that the design of the current layer setup came about through trial and error. There were no pre-written rules as to which layers, or how many layers to use, but only successful examples from other cases. From this perspective, the engineers' role has shifted from making a perfect predictive model to creating an imperfect intelligent agent that will later find its own control strategies. In a model-centric paradigm, such as MPC, the engineers study the system to be controlled, and seek optimal control policies. In a non-model-centric paradigm, such as RL, however, engineers distance themselves from the real problem at hand and relinquish their control, placing their hope in a flawed AI agent that interprets their instructions and accomplishes their plans.

This is not the only instance in which engineers must choose how the agent should be built. During a lab meeting, Bowes reported that overfitting had occurred in training the RL agent. Overfitting describes a phenomenon in machine learning whereby agents exhibit poorer performances when trained for too many episodes. We may understand a machine-learning problem as a curve-fitting problem, attempting to fit a curve to data points in a coordinate system in order to describe their general distribution pattern. We may draw a simple curve that touches no data points in the coordinate plane, or we may draw a complex curve to connect all the points. The first curve is not optimized, whereas the latter accurately describes each data point. However, this "overfitted" curve fails to describe the general pattern in the dataset; whenever a new instance arrives, it falls outside the curve. Overfitting also occurred in Bowes's RL experiment, where the agent was exposed to limited storm events to learn control policies. Training the agent for too many episodes results in an overfitted agent that performs poorly when facing new storm events. The engineer must observe an agent's behavior to ascertain a "sweet spot" at which a training session must end to avoid overfitting.

With the rise of data-driven approaches comes the emergence of intelligent agents. When these AI agents manage diverse environmental processes, concerns regarding control shift from model accuracy to a sense of trust. The questions which engineers need to consider are no longer about control but relinquishing control. We are aware that AI agents can produce desired outcomes or even out-perform human operators, but are we comfortable and confident enough to upload these agents into cyberphysical systems and allow them to run the actuators? What is our tolerance for relinquishing human control? How do we frame unexpected outcomes produced by these AI agents? Let us use another, more extreme, example as an analogy to examine what is at stake. Considering that all new Tesla cars arrive with autopilot features and full self-driving capabilities, self-driving cars are a reality. The algorithms and hardware of a self-driving car out-perform human drivers on many levels. A self-driving car possesses better sensors, reacts faster, never gets tired, and obeys traffic rules better than most human drivers. However, how comfortable would you feel in a self-driving car at 70 miles per hour on a crowded freeway, even though you know its driving out-performs yours?

These questions simply cannot be answered within the epistemic framework which early cyberneticians used to conceptualize control and agency from an anthropocentric perspective. To entertain these questions, we must employ a posthumanist, co-production framework, as described in Chapter 8, conceptualizing intelligence as distributed capacity in a network of human and nonhuman agents, including these intelligent machines.

Choreographing Intelligent Machines, and Halprin's Scores

We must again employ a design research framework to consider the above engineering cases. Within a mainstream engineering framework, the question is whether we trust these agents, and how to make them more trustworthy. From this perspective, certain machine behaviors become unwanted, like opening and closing valves multiple times to cause insignificant downstream floods. Design, however, allows us to connect thinking from one field to another; we can situate these engineering examples in a posthumanist framework, translating them into conceptual strategies to intervene in the cybernetic environment on a systemic scale.

Chapter 8 reviewed a range of RL agents, such as AlphaStar and AlphaGo, that developed strategies of their own, surprising human players. Based on these examples, landscape architects and ecologists have speculated on the implications of machine learning in environmental management. In a 2016 paper called "Designing Autonomy: Opportunities for New Wildness in the Anthropocene," landscape architect Bradley Cantrell and his collaborators imagined an RL-based AI system called "wildness creator," which devises its own strategies and creates wild places beyond human comprehension.

The paper questions the notion of wilderness and provides an alternative way to conceptualize the use of machines in constructing wild places.[18] It is often believed that machines, being artificial, are inherently at odds with the notion of wild, which is on the line of nature. However, this reasoning places moral limitations regarding which strategies are appropriate, based on the standards of those with the privilege and authority to define "nature" and "wild." Ironically, if one could construct wild places, those places would no longer be wild, because their construction would be based on human models and comprehensible by humans; if we are able to predict everything in a system, nothing wild remains. However, the "wildness creator" renders a completely different scenario, in which machines begin to devise strategies based on their own observation and calculation. Places constructed by these machines are truly wild—they are novel ecologies for which humans have no model.

This thought experiment may have appeared outlandish in 2016, but when we consider the above-described work carried out by the Link Lab, especially that of the RL agent, a "wildness creator" seems within reach. In order to characterize machine strategies, we will continue to use this example as an object of analysis. In the RL example, the agent is trained to prevent floods. Yet landscape architects and ecologists often advocate for the importance of flood events for ecosystems. One might imagine an RL agent trained to strategically promote flood events in certain areas of a city, creating novel and emergent ecologies. Moreover, the RL agent may actively maintain water depths in the retention ponds; thus, one might also deploy the agent to manage these bodies of water and provide ideal conditions for wetland species. However, these considerations do not speak to the notion of *wildness,* since they are, in essence, objective optimization problems articulated with reward functions. In the end, the AI agents are still considered goal-seeking machines attempting to maximize objective functions defined by engineers.

To truly consider the meaning of constructing wild places, we must focus on situations where outcomes produce side effects beyond the articulated goals. Scientists from institutions such as Google Brain and OpenAI have proposed five concrete problems in AI safety; the first is "avoiding negative side effects."[19] This problem is profoundly related to reward function and goal articulation. In RL, the agent attempts to optimize the reward function through which scientists articulate their intended goals. However, goals articulated by scientists are not inevitably reflected in desired agent behavior. When considering AI reward functions, we must realize that in assigning a reward value to the desired outcome, we are also assigning zero values to all other possible outcomes. An agent ignores whatever is not articulated in the reward function. To use a thought experiment in AI safety research, let us consider a cleaning robot rewarded for cleaning a room. We assign values to each bit of refuse the robot picks up. However, how do we ensure the robot will not knock over a vase simply because doing so will speed up the cleaning task? The vase is assigned zero value in the objective function, so that the cleaning robot will not attend to *not* damaging it.[20]

This is not purely a thought experiment, since the AI community confronts similar problems daily. Let us extend the stormwater RL example in the Link Lab research, and we will find room to consider how the reward function may produce unexpected ecological consequences. Recall its reward function:

$$r = \begin{cases} -\sum(\text{flooding}), F > \delta \\ -J1_{\text{flooding}} - (|St1_{\text{depth}} - \text{target}| + |St2_{\text{depth}} - \text{target}|), F = 0 \end{cases}$$

The benefit of the second part of the function is that, right after a rainfall event, the agent would open valves to quickly discharge water, to maintain the desired water depth in the retention ponds. Yet, as discussed, this function has an inherent bias. Flooding is characterized by overall volume rather than frequency. In other words, the frequency of flood events is assigned zero value; the agent ignores it and will even exploit it. The agent performed this exact ploy in order to "game" the reward function. It would alternately open and close the valves to discharge water right after the rain, even though this strategy produced multiple small flood events at the downstream J1 node. If we imagine ourselves as residents living beside J1 in this simulated SWMM environment, we may observe a fascinating phenomenon. After an upstream rain event, the J1 node will experience mysterious repeated flooding, which will drastically alter the local ecology. Thus, the ecology in J1 is de-coupled from a local weather pattern, but coupled with a weather pattern elsewhere. Landscape designers would see opportunities to design with this behavior, to de-couple and re-couple landscape systems. This displacement of phenomena creates unusual experiences, perhaps providing another way for urban residents to experience ecosystem dynamics. We could provide a plausible explanation in this case by analyzing the reward function, but the RL agent devised other strategies that are difficult to explain. For example, it would occasionally keep the valve half-open, an apparently unnecessary action. How would these unexpected strategies shift ecosystem dynamics around the stormwater system?

Finally, considering the entire drainage system, the RL agent essentially produces a new hydrological pattern. Over time, new ecologies would consequently emerge and be managed by another AI agent. The result: a constructed wild place highly maintained by AI agents whose actions may exceed human comprehension. Oddly enough, to express our needs, rather than directly interacting with the environment, we communicate with AI agents through reward functions.

Since 2021, the dMIST scientists have made several updates.[21] First, instead of using SWMM as a toy system to replace the real environment, the team selected a testing site in the City of Norfolk, with real-world information collected by sensors across the system. Second, aside from flooding, the team incorporated another goal in the objective function—to mitigate water pollution by holding water in the retention pond long enough to encourage sedimentation. The AI agent must now consider three objectives: overall flooding in the system, target levels in the retention pond, and pollution indicators downstream. Moreover, because this is a real site, different nodes exhibit different conditions. Based on historical data, several are known to be prone to flooding, while others may be able to withstand minor floods. In order to address this issue, the team gave different weights to different nodes in the reward function. The hope is that the agent would sacrifice unimportant nodes to prevent floods in the system.

The agent now must consider multiple conflicting goals: if the agent holds water in the ponds for too long, that may cause upstream floods. If it releases water too quickly, flooding would occur downstream, and pollutants would overwhelm the stormwater system. Moreover, the agent must also consider different node conditions, and treat them separately. The dMIST scientists compared the simulation result with rule-based control, optimized for water quality, and the results raise interesting questions. In rule-based control, the water is held long enough to let sediments drop, and downstream pollution is thus significantly reduced. Because the RL agent must consider other objectives, it cannot compete with a rule-based control strategy. During a weekly meeting, one additional piece of information was presented to the dMIST team: the City of Virginia Beach had implemented a system built to optimize water pollution reduction, but in some nodes, this system caused even more flooding. In other words, a one-size-fits-all strategy might not work in some nodes.

These RL experiments reveal an entirely different venue to consider a flexible framework that accounts for site-specific goals. Rule-based control applies a one-size-fits-all solution to a problem requiring diverse strategies. If we contrast flooding mitigation and pollution control as two ends of a spectrum, we may now imagine a family of RL agents scattered along this spectrum, each of which devises a slightly different control policy. These agents could be deployed in discrete sub-systems requiring diverse treatment. The environment is a complex of many more than two objectives and goals—it is a high-dimensional space with a limitless number of potential variables. These variables could provide a high-dimensional objective function to train a variety of agents that each devise slightly different strategies. If these agents were deployed within the system, their interactions would produce further emergent behaviors. There thus exists a viable technical framework based on RL to produce *wildness*, cultivated by distributed intelligent agents.

At this point, many would counter-argue that the RL agent has been trained and tested in a simulated environment built with SWMM, and, therefore, it would perform poorly in the physical environment. However, RL is non-model-centric; this means that the RL agent does not rely on an internal model of the system to compute control policies. Instead, an RL agent itself is a set of policies. When the agent is uploaded to a real system, it is coupled with the system using only the sensors distributed throughout the system. The sensing network will produce a state-space representation, with state variables, that will be identical to that offered by the SWMM simulation. In other words, in the SWMM model, the agent observes the system through state variables, including water depth for each pond and node, rain forecasts, tidal information, and valve states. When uploaded to a physical system, the RL agent knows nothing more. For the RL agent, the SWMM simulation and the physical system appear as two identical matrixes of state variables. From the perspective of autopoiesis, the RL agent only interacts with a reality reconstructed by its sensing peripheries. Thus, the SWMM simulation is as real as the physical environment.

Let us now turn to landscape design, and rely on the notion of choreography to further illustrate the operation of a designer relying on intelligent machines in a cybernetic environment. In the 1970s, landscape architect Lawrence Halprin introduced the RSVP cycle: Resources, Score, Valuaction, and Performance. Together, they describe the creative processes in design. Resources entail anything used in the creative process, including time, physical materials, other people, ideas, and limitations. Score is likened to musical scores, a series of instructions for the work. Valuaction, a term coined by Halprin, is a process of dynamically responding to the creative work based on personal and professional values. Performance stands for setting the creative work in motion.

Lawrence Halprin developed RSVP with his wife, Anna Halprin, a choreographer influenced by the open-score movement in music and choreography in the 1960s. At the time, many musicians and choreographers embraced participatory and open-ended compositions and choreographies, known as "open scores," which replaced the traditional notational systems that decisively directed performers.[22] Open scores embraced instead interpretative chance and flexibility by giving visually evocative approximations, suggesting general intentions and interpretation paths. In the 1960s, Lawrence Halprin was interested in community design that sought to incorporate local desires as part of public landscape projects. In exploring community design processes, the Halprins collaborated on experimental performance projects, using open-scoring ideas.[23] When introducing the RSVP cycle, Halprin used choreography examples several times in his book. The Halprins' collaboration, and the influence of choreography, became facts which every landscape architect learned in school. Due in part to this history, the notion of choreography was distilled into the landscape discourse.[24]

Today, landscape architects deploy this dance term in describing their works. Brian Davis, for example, introduces his "instrument tables" that allow designers "to *choreograph* interactions and patterns [of different instruments or agents] on a formal level."[25] Similarly, Cantrell and Holzman used "choreograph" throughout their book *Responsive Landscapes* when introducing responsive frameworks for landscape design.[26] In landscape architecture, the term *choreograph* speaks to a sense of relationship formed between landscape designers and other intelligent agents and systems—nonhuman species, environmental processes, cultural practices, land use policies—that shape the environment. Designers cannot authoritatively control these agents to produce landscapes according to a well-articulated plan. Instead, designers rely on "open scores" to influence agents to act within a range of intentionality. The agents are at liberty to interpret, and the outcomes are filled with uncertainty. To a certain extent, contemporary landscape design is a practice of exploring different choreography techniques.

In the examples above, applying the notion of choreography to reframe the relationship between the engineers and the RL agents, we could find diverse analogies. Engineers communicate with the RL algorithm only with objective functions, which, as we saw, leave room for uncertainty, and thus flexibility. In a non-model-centric paradigm, where we use machine learning to build intelligent models to carry out our intentions, the descriptions of our intentions become "open scores" that encourage uncertainties. Therefore, in a non-model-centric paradigm, we need a framework that allows us to embrace uncertainty, viewing it as possibility rather than as a threat to controlled stability. Attuning, then, becomes the concept of reframing the relationship among the agents in this new framework.

Notes

1 Metcalf & Eddy, Inc., University of Florida, and Water Resources Engineers, Inc., "Storm Water Management Model Volume I - Final Report."
2 Rossman, "Storm Water Management Model Reference Manual Volume II – Hydraulics."
3 https://friendsofmrbm.org/model-history. Last accessed: April 29, 2024.
4 I owe this idea to Alexandra Dimitri, a landscape architect who proposed to treat street puddles as sites for water retention units and emergent urban ecosystems. See https://www.asla.org/2015studentawards/102410.html to learn about the design proposal.
5 Winsberg, *Science in the Age of Computer Simulation*, 45.
6 Winsberg, 45.
7 Winsberg, 45.
8 Sadler et al., "Exploring Real-Time Control of Stormwater Systems for Mitigating Flood Risk Due to Sea Level Rise."
9 Loftis et al., "StormSense."
10 Sadler et al., "Leveraging Open Source Software and Parallel Computing for Model Predictive Control Simulation of Urban Drainage Systems Using EPA-SWMM5 and Python"; Sadler et al., "Exploring Real-Time Control of Stormwater Systems for Mitigating Flood Risk Due to Sea Level Rise."

11 Bowes et al., "Flood Mitigation in Coastal Urban Catchments Using Real-Time Stormwater Infrastructure Control and Reinforcement Learning."
12 Brunton and Kutz, *Data-Driven Science and Engineering*.
13 Bowes et al., "Flood Mitigation in Coastal Urban Catchments Using Real-Time Stormwater Infrastructure Control and Reinforcement Learning."
14 Bowes et al.; Saliba et al., "Deep Reinforcement Learning with Uncertain Data for Real-Time Stormwater System Control and Flood Mitigation."
15 Lillicrap et al., "Continuous Control with Deep Reinforcement Learning."
16 Saliba et al., "Deep Reinforcement Learning with Uncertain Data for Real-Time Stormwater System Control and Flood Mitigation"; Bowes et al., "Flood Mitigation in Coastal Urban Catchments Using Real-Time Stormwater Infrastructure Control and Reinforcement Learning."
17 Bowes et al., "Flood Mitigation in Coastal Urban Catchments Using Real-Time Stormwater Infrastructure Control and Reinforcement Learning."
18 Cantrell, Martin, and Ellis, "Designing Autonomy."
19 Amodei et al., "Concrete Problems in AI Safety."
20 Amodei et al.
21 These preliminary outcomes were reported by Ben Bowes during weekly meetings in February 2021.
22 John Cage was one of the leading compositional figures in this avant-garde movement. His famous composition 4′33″ may be understood as an attempt to embrace flexibility in interpretation. The piece consists of blank music sheets running 4 minutes and 33 seconds long. They may be "played" with any instrument or combinations of instruments.
23 Lystra, "McHarg's Entropy, Halprin's Chance," 77.
24 There is little historical research as to how "choreograph" attained its currency in today's landscape design.
25 Davis, "Landscapes and Instruments," 302. Emphasis added.
26 Carnell and Holzman,. *Responsive Landscapes: Strategies for Responsive Technologies in Landscape Architecture.*

Bibliography

Amodei, Dario, Chris Olah, Jacob Steinhardt, Paul Christiano, John Schulman, and Dan Mané. "Concrete Problems in AI Safety." *arXiv Preprint arXiv:1606.06565*, 2016. https://arxiv.org/abs/1606.06565.

Bowes, Benjamin D., Arash Tavakoli, Cheng Wang, Arsalan Heydarian, Madhur Behl, Peter A. Beling, and Jonathan L. Goodall. "Flood Mitigation in Coastal Urban Catchments Using Real-Time Stormwater Infrastructure Control and Reinforcement Learning." *Journal of Hydroinformatics*, 2020. https://doi.org/10.2166/hydro.2020.080.

Brunton, Steven L., and Jose Nathan Kutz. *Data-Driven Science and Engineering*. Machine Learning, Dynamical Systems, and Control. Cambridge: Cambridge University Press, 2019.

Cantrell, Bradley, and Justine Holzman. *Responsive Landscapes: Strategies for Responsive Technologies in Landscape Architecture*. New York: Routledge, 2015.

Cantrell, Bradley, Laura J. Martin, and Erle C. Ellis. "Designing Autonomy: Opportunities for New Wildness in the Anthropocene." *Trends in Ecology & Evolution* 32, no. 3 (2017): 156–66. https://doi.org/10.1016/j.tree.2016.12.004.

Davis, Brian. "Landscapes and Instruments." *Landscape Journal* 32, no. 2 (2013): 293–308.

Lillicrap, Timothy P., Jonathan J. Hunt, Alexander Pritzel, Nicolas Heess, Tom Erez, Yuval Tassa, David Silver, and Daan Wierstra. "Continuous Control with Deep Reinforcement Learning." *arXiv:1509.02971 [Cs, Stat]*, July 5, 2019. http://arxiv.org/abs/1509.02971.

Loftis, Jon Derek, David Forrest, Sridhar Katragadda, Kyle Spencer, Tammie Organski, Cuong Nguyen, and Sokwoo Rhee. "StormSense: A New Integrated Network of IoT Water Level Sensors in the Smart Cities of Hampton Roads, VA." *Marine Technology Society Journal* 52, no. 2 (March 1, 2018): 56–67. https://doi.org/10.4031/MTSJ.52.2.7.

Lystra, Margot. "McHarg's Entropy, Halprin's Chance: Representations of Cybernetic Change in 1960s Landscape Architecture." *Studies in the History of Gardens & Designed Landscapes* 34, no. 1 (January 2, 2014): 71–84. https://doi.org/10.1080/14601176.2013.850313.

Metcalf & Eddy, Inc., University of Florida, and Water Resources Engineers, Inc. "Storm Water Management Model Volume I - Final Report," July 1971.

Rossman, Lewis A. "Storm Water Management Model Reference Manual Volume II – Hydraulics." Washington, DC.: U.S. Environmental Protection Agency, May 2017.

Sadler, Jeffrey M., Jonathan L. Goodall, Madhur Behl, Benjamin D. Bowes, and Mohamed M. Morsy. "Exploring Real-Time Control of Stormwater Systems for Mitigating Flood Risk Due to Sea Level Rise." *Journal of Hydrology* 583 (April 1, 2020): 124571. https://doi.org/10.1016/j.jhydrol.2020.124571.

Sadler, Jeffrey M., Jonathan L. Goodall, Madhur Behl, and Mohamed M. Morsy. "Leveraging Open Source Software and Parallel Computing for Model Predictive Control Simulation of Urban Drainage Systems Using EPA-SWMM5 and Python." In *New Trends in Urban Drainage Modelling*, edited by Giorgio Mannina, 988–92. Springer International Publishing, 2019. https://doi.org/10.1007/978-3-319-99867-1_170.

Saliba, Sami M., Benjamin D. Bowes, Stephen Adams, Peter A. Beling, and Jonathan L. Goodall. "Deep Reinforcement Learning with Uncertain Data for Real-Time Stormwater System Control and Flood Mitigation." *Water* 12, no. 11 (November 2020): 3222. https://doi.org/10.3390/w12113222.

Winsberg, Eric. *Science in the Age of Computer Simulation*. Chicago, IL: University of Chicago Press, 2010.

13

ACTUATING LEADS TO ATTUNING

Cultivated Wildness

Before venturing into the last category of practices, let us reflect on what we have explored so far. The feedback control diagram points to two different paradigms. The first one is a model-centric paradigm built around sensing-computing-actuating feedback mechanisms. A collection of processes may be grouped as a dynamic system with measurable inputs and outputs. Sensors are added to collect data that feed back or feed forward to the controller to compute actuating strategies. With big data, we might anticipate more learned state-space models of the environment. These state-space models are faster to compute and simpler to optimize and are used to bypass physics-based empirical models, such as SWMM, for environmental management. Ever-increasing computing power will make it feasible to run live parallel simulations of the environmental systems in order to devise control policies, evaluate them based on preset goals, and adjust strategies on the run. Sensing networks, big data, state-space modeling, and machine-learning techniques set up a technical foundation for applying modern control strategies, such as model predictive control (MPC) algorithms, for managing the environment. In this mainstream guide toward a version of the cybernetic environment, we anticipate numerous models distributed throughout the environment, continuously forecasting future scenarios based on sensed data, devising live management strategies through actuators, and reevaluating these strategies based on sensed data returned to the models. We could anticipate improved controls over all sorts of environmental systems to urge them in the desired direction—more resilient, more adaptive toward climate change, and with an enhanced response to disturbances. This framework is a type of adaptation based on control: more information leads to better understanding and modeling of the environmental systems, and thus better control regimes toward a climate-resilient future.

DOI: 10.4324/9781003320852-17

The second paradigm points in another direction, that of a sense of cultivated wildness. Tracing a range of contemporary data-driven practices in art and engineering demonstrates that applying cybernetic thinking to environmental practices may lead to something other than control. Many contemporary art and design practices present ways to "misuse" sensing technologies. Here, sensing is more than data collection, but it could be framed as coding new state spaces by which the environment is revealed to us in ways we have not yet imagined. Meanwhile, the application of deep reinforcement learning (RL) in environmental management sheds light on a non-model-centric paradigm with distributed AI agents that devise unexpected policies. These developments point to a version of cybernetic thinking that is not solely about control. Instead, we encounter uncontrollability and wildness.

Further questions arise: if repeated actuating leads to wildness, then what has been actuated? If actuating is detached from humans' exercise of agency, how might we even conceptualize it? Answering these questions requires design thinking to connect ideas across different fields, bringing them within a posthumanist co-productive framework. Only in this way will we begin to unpack the notion of cultivated wildness.

Here, we need the concept of attuning in order to proceed. Attunement was introduced by object-oriented ontology (OOO) proponent Timothy Morton:

> Since a thing can't be known directly or totally, one can only attune to it, with greater or lesser degrees of intimacy...Since appearance can't be peeled decisively from the reality of a thing, attunement is a living, dynamic relation with another being.[1]

Morton uses the example of an opera singer's voice so attuned to the frequency of a wine glass, it shatters. Within the posthumanist co-productive framework, attunement is where intelligence emerges. Machines become necessary media for designers to cultivate attunement between assemblages. Repeated actuating becomes a type of living, dynamic relation between machines and other assemblages. The following attuning examples can be explored through two spatial scales. We can explore distributed intelligence and responsive frameworks on a landscape scale or on a macroscale. Then, we will zoom in to a garden, observing how individual machines such as robots interact with nonhuman agents such as plants.

Macroscale: Distributed Intelligence and Responsive Frameworks

We tend to ignore how widely intelligent machines are already distributed in the environment, overlooking the scale on which they participate in many environmental processes. For example, the Everglades restoration project is considered an exemplar of the adaptive management paradigm, a decision-making method emphasizing learning by doing, achieved through intense monitoring and actuating processes. Established in 1949, the South Florida Water Management District (SFWMD) is a regional governmental agency that manages water resources in the southern half of Florida, including the Everglades. This 1.5-million-acre wetland sustains numerous wildlife species—the ultimate "wild place" in the eyes of urban residents. However, the "wild" Everglades is, in fact, a highly maintained place. In South Florida, numerous sensing stations are installed across bodies of water, generating real-time hydrological and water quality data for building simulation models, such as the South Florida Water Management Model. Local organizations and agencies use these models to analyze operational changes to their water systems and make informed management decisions. Moreover, thousands of miles of engineered canals and pipes are carved into and buried under Florida's landscape, with water control infrastructures—including water basins, spillways, weir gates, pumps, dams, and locks—strategically placed along the waterways. These actuators within the system directly influence the hydrological patterns of South Florida. Thus, both the amount of water and hydrological patterns are carefully calculated and controlled using simulation models based on real-time hydrological data, weather forecasts, climate modeling, and historical data.

This South Florida project is only one instance of numerous efforts to use cyberphysical systems to manage environmental processes. Recent discourse regarding smart cities has fueled cybernetic imaginings across social sectors. The environment is already laden with intelligent machines waiting to be mobilized by designers and engineers, and we can only anticipate their increase. Yet the ultimate question for the next generation of designers becomes how to develop a framework to communicate and design with these machines, transforming them into actuators to cultivate wildness. Two speculative landscape research projects, supported by robotic and machine-learning examples, shed light on an adaptive framework based on real-time feedback and machine learning.

In the first project, "Towards Sentience™," designer Leif Estrada proposed to distribute intelligent machines and sensing networks throughout the Los Angeles River in order to influence hydrological patterns and, over time, create land.[2] Estrada tested the sensing-processing-actuating responsive framework on a hydromorphology table at the Responsive Environments & Artifacts Lab of Harvard University's Graduate School of Design. The sediments and water flow inputs were controlled through four material feeders and a water pump, to simulate fluvial sedimentation behaviors. The table was equipped with sensing and monitoring devices to gather real-time data, including a Microsoft Kinect above the table and ultrasonic sensors downstream. Real-time data was then fed into the 3D modeling software Rhinoceros3D through Grasshopper, a visual programming plug-in. In one of his design experiments, Estrada proposed an actuating system called "Attuner," which consisted of a matrix of servo motors hanging above the table. Each motor drove an acrylic dowel, and the bottom portion of the dowel projected into the water. The turning of the servo motors drove the dowels up and down to influence the flow pattern, which could create different landforms downstream. The topography was live-tracked by sensors, and elevation data were used to build a digital elevation model to identify a series of highlands and lowlands. This dataset was then used to inform operations, such as adding to a highland by depositing sand upon it, or eroding highlands by directing more water toward them.[3] Estrada reported that his cyborg system demonstrated a level of live updates and feedback beyond human capacity.[4]

Estrada's experiment tested the sensing-processing-actuating feedback loop as a viable responsive landscape framework, utilizing sensors and actuators in the environment for deploying real-time landscape strategies. If we extend this experiment, we could imagine a landscape laden with sensors and actuators, akin to Estrada's experiment, serving as a basis for RL agents to learn different policies. These RL agents would be deeply embedded in the environment, which over time would co-evolve with human and nonhuman agents.

This type of speculation raises two concerns. First, continual learning strategies are needed to anticipate an intelligent machine that co-evolves with human and nonhuman agents in the environment. Most AI algorithms are considered "offline," meaning that, after training, what is uploaded to a cyberphysical system is a fine-tuned neural network. Training processes tune the weights in each neuron by exposing a neural network to large amounts of data. The neural network is calibrated to its optimal condition when training is finished, and when in operation, it simply generates outputs based on input information, without learning anything new. If we consider the effects of evolution and climate change, an agent will quickly grow outdated. Consequently, developing a continual learning framework is the next challenge in incorporating intelligent machines for managing the environment. Second, machine learning is often divided into training and testing sessions. Training data is available for machines to learn from before the agent is applied in real-world situations. In RL, an agent develops its strategies within a simulated environment, such as a video game played a limitless number of times. However, the environment possesses no reset button, and if an agent caused an unwanted result, there would be no way to undo its effects. This poses ethical challenges in applying AI agents to environmental management.

Another speculative research project provides insights into these predicaments. In his MIT architectural thesis research, designer Ricardo Jnani Gonzalez proposed a system deeply embedded in a cryosphere environment.[5] The system consists of a central "mind," which can be understood as a supercomputer unit, and distributed "bodies," actuators that alter the physical environment. In operation, the "mind" first casts a vast array of actuating policies across the "bodies." Different local environments respond to these policies in separate ways. The "mind" then evaluates the policies based on discrepancies between projected scenarios and actual outcomes. Then, the "mind" updates the policies and returns them to the "bodies." If one policy yields the best outcome, its successful experience is embedded within the next iteration of intervention. Thus, one "body" might influence other "bodies" by transferring its knowledge. Through this iterative process, the distributed system could gradually attune to and evolve with the cryosphere environment.

Although speculative, this project hints at a technical framework that mirrors two widely used machine-learning algorithms, genetic algorithms and transfer learning. A genetic algorithm was used in the Link Lab MPC example described in Chapter 12. Genetic algorithms are inspired by the natural selection process, in which the best qualities of one generation pass to the next generation. In each iteration of training, an array of policies is deployed and the best are retained. In the next generation, mutations are introduced to successful policies, and a new array of updated policies is deployed. Repeating this selection-mutation process over time may increase the algorithms' performance. For environmental management, we should not expect a single complex model to compute one-shot policies. Instead, we could imagine distributed agents, each taking incremental interventions, sharing knowledge with each other, and evolving together. Research already uses robots to deploy incremental and small-scale interventions to manage the environment or conduct construction. Romu is a robot designed to drive interlocking sheet piles into the ground to build check dams in arid regions, which help prevent erosion and promote groundwater recharge.[6] Introduced earlier in the book, RangerBot is a vision-based underwater robot that identifies and kills coral-eating starfish, and monitors reef health and water quality to protect Australia's Great Barrier Reef. These robots could act as distributed "bodies" with which machine-learning agents evolve and develop strategies.

In Gonzalez's speculation, a body can transfer successful policies to other bodies. This involves transfer learning, in which knowledge acquired from one domain is applied to another. For example, ChatGPT, a web chatbot that gained popularity at the end of 2022, has been integrated into different fields. Here, GPT stands for "Generative Pre-trained Transformer," which is essentially a language processing model that uses deep learning to generate human-like texts.[7] ChatGPT is a variant of GPT specifically designed for chatbot applications. However, because ChatGPT was trained with numerous texts from the internet, including lines of computer codes, it started to pick up skills such as coding. Users found that GPT can generate SVG plots and write HTML code to generate web layouts from simple written descriptions. People quickly discovered that ChatGPT could perform many other tasks, from writing computer code to copyediting text. ChatGPT now has an API (application programming interface) that allows users to employ GPT builders to customize their GPTs for specific purposes, like "making a software engineer who helps format my code."[8] It even allows users to upload their files to include other knowledge bases, to further tweak their GPTs. Few-shot learning is an approach in machine learning that enables a pre-trained model to generalize over new categories of data, thus tuning the model to specific knowledge domains.

In other examples, scientists explored strategies with which to transfer policies built for one type of robot to another with vastly different hardware properties,[9] or to generalize and quickly transfer policies from various robotic systems to a new robot.[10] With these concrete developments in AI research, we may speculate that transfer learning could provide technical possibilities for Gonzalez's proposal, in which one actuator's experience may be generalized and transferred to another. Agents pre-trained with general knowledge about the system before being distributed in the environment could rapidly attune to different conditions, and successful experiences could be generalized and applied to other tasks.

At this point, it should be clear to readers that many speculative design research projects, such as the ones presented above, are grounded in technologies currently in development rather than mere speculation. They may sound outlandish at first hearing, but they provide alternative paths for future AI advancement. In Social Construction of Technology (SCOT), "stabilization" describes the process of the social construction of a technological artifact, and the result may be a "technological frame," which "structures the interactions among the members of a relevant social group, and shapes their thinking and acting."[11] Analyzing AI development with a SCOT lens, I argue that society is in the midst of constructing a technological frame for AI agents. Thus, it is critical for different social groups, including designers and environmentalists, to imbue their values and concerns into a frame that will eventually guide how AI agents are incorporated into environmental management. From this vantage, the value of these speculative projects is in challenging computer scientists and engineers to reconsider the ramifications of AI technology outside a control and optimization framework. These projects present ways to reimagine AI agents as distributed co-creators, to cultivate wild places, and to support emergent ecologies with endless potential.

Microscale: Robots in the Garden

How might we cultivate wildness by developing tools for direct interaction with nonhumans? In this case, we will use plants as an example; two models exist for this task, one used in agriculture practices, the other in art.

Many contemporary farms have begun incorporating machines for precision farming, which uses sensors to monitor growing conditions and robotics to fertilize, water, and harvest crops at an individual plant level. Precision farming aims to manipulate plants for greater productivity by creating a single variant system that optimizes only one aspect of plant life. It thus increases productivity by eliminating "wildness" in plants; plants are exploited for human use.

The second model is found in art. David Bowen uses cybernetic technologies to create art installations that explore the relationship between machines and plant materials (Figure 13.1). "Growth Rendering Device" was a feedback system between machine and plant.[12] In a bottle attached to a vertical scanner, a pea plant was suspended in a hydroponic solution. This setup was attached to an inkjet printer head over a roll of paper mounted on the wall. The system recorded the plant's growth by producing drawings based on the scanned data every 24 hours. The system then scrolled the roll of paper and began the next drawing cycle. As Bowen suggested, the outcome was not predetermined; the device might record growth, but also, likely, the decay and demise of the plant. In a similar fashion, Bowen's "Growth Modeling Device"

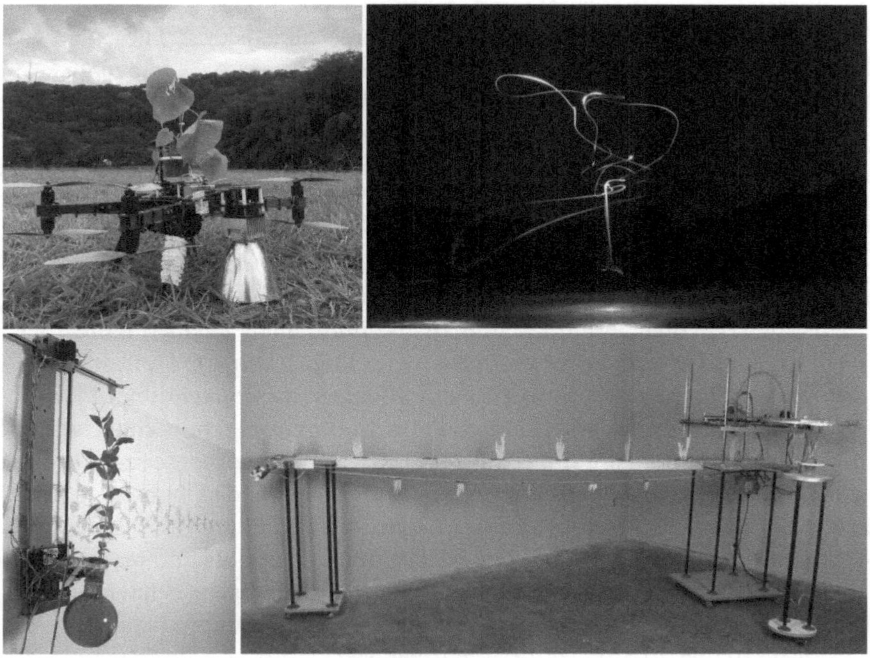

FIGURE 13.1 Top: "Plant Drone" (2019); bottom left: "Growth Rendering Device" (2007); bottom right: "Growth Modelling Device" (2009). Courtesy of David Bowen.

scanned a plant from three different angles and 3D-printed the results over the course of 24 hours.[13] A conveyor belt advanced approximately 17 inches after each printing cycle to begin the next scan-print cycle. Bowen's "Plant Drone" project built a plant pilot.[14] He mounted a plant onto a drone and attached electrodes to its leaves. The bioelectricity emitted by the plant leaves determined the drone's movements. The plant "piloted" the drone under the night sky, its movements captured with long-exposure photography.

Although art production is commonly assumed to be an entirely human endeavor, in these installations, the drawings, models, and photographs are completely determined neither by the artist nor by the machines; instead, they are co-produced by machines and plants in a collaboration enabled by the artist. Bowen built unconventional machines that gave plants agency in art production.

These examples demonstrate that it is possible to develop cybernetic systems that bypass control and optimization. With this conviction, I established the Ecological Laboratory for Urban Agriculture (ELUA) in 2021 at the City College of New York, as a prototyping platform to explore co-productive intelligence among humans, robots, and plants. Quickly, ELUA installed robots from FarmBot, a California-based firm that designs and markets open-source commercial gardening robots and develops web applications to use them. These are gantry robots that operate in three dimensions. The robot arm can pick up different tool heads to rake soil, plant seeds, water plants, and eliminate weeds. Users can also design and 3D-print their own tool heads to perform other tasks. Furthermore, ELUA's interdisciplinary team comprises landscape architects, architects, and computer scientists, who will use machine-learning techniques to train the robots to cultivate a polyculture food garden. We hope the robots will devise interesting gardening strategies that surprise and inspire their human counterparts. Here, robots are regarded as important agents in urban ecosystems, and, in the face of climate change, can adapt and co-evolve with other human and nonhuman agents (Figure 13.2).

Despite ELUA's practical concerns regarding urban food production, it proposes a thought experiment that challenges our conventional views on agency and intelligence and questions how the environment is conceived and constructed. It offers a glimpse into an ecosystem of computerized ecology, where the relationship between humans and plants is deeply mediated and, at the same time, enabled by sensors, controllers, computer hardware, layers of computer code, and online servers. The humans, robots and plants form a positive feedback loop that produces spin-off and emergent behaviors. Thus, the outcome is a cultivated wildness—unexpected and unpredictable.

Conceptually, ELUA lies between engineering approaches, such as precision farming, and artistic expressions, such as Bowen's installations. The project is not an industrial prototype to be monetized as a product. Similar systems are already in practice, especially in the fields of precision farming

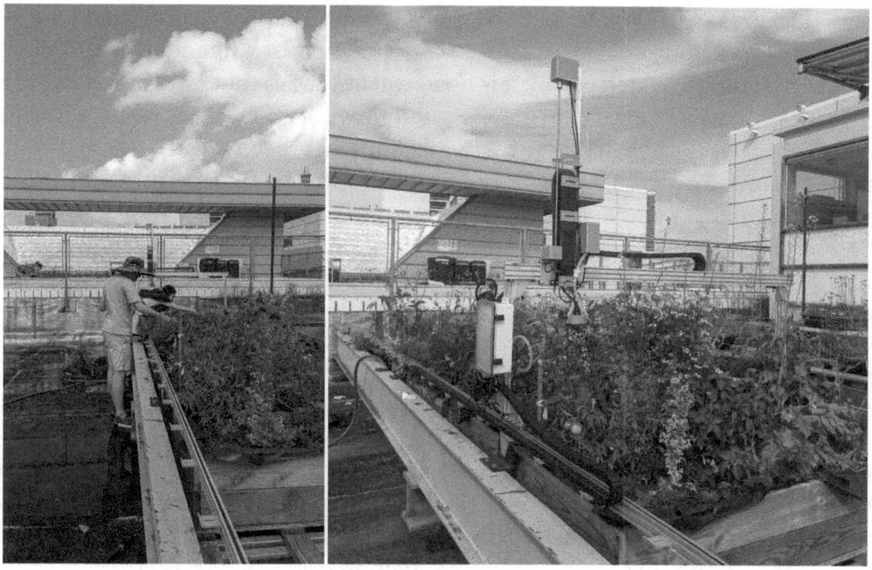

FIGURE 13.2 Ecological Laboratory for Urban Agriculture (ELUA) at the City College of New York. Photos by the author.

and robotic agriculture. However, unlike artistic expressions that function at a conceptual level, ELUA is an ongoing process with a pragmatic goal. The project establishes a platform for designers to test landscape strategies focusing on interactions between machines and plants. Moreover, ELUA calls for further explorations between art and science. The technique used in this installation can be conceptualized as post-prototyping. It denaturalizes mainstream practices, reframing them in a different discourse to cultivate alternative understandings and reveal unexplored territories.

Cybernetics without Control

The emerging practices in cybernetics challenge the idea that communication inevitably leads to control, and that recursive processes always lead to stability. The implications of communication without control, and control without stability, raise questions about how we define and conceptualize these ideas themselves. To explore cybernetics beyond control, we need to rely on the oxymoron "cultivated wildness." This term seems contradictory. How can we cultivate something that implies a lack of human control? This tension establishes an unfamiliar territory in which to entertain cybernetics beyond control.

Before we proceed, it is important to address several caveats. As demonstrated in Chapter 4, the concept of wildness cannot be separated from that of wilderness. Environmental historian William Cronon notes that "modern environmental movement is itself a grandchild of romanticism and post-frontier ideology, which is why it is no accident that so much environmentalist discourse takes its bearings from the wilderness these intellectual movements helped create."[15] The image of a wild place has become intertwined with moral values and cultural symbols deeply rooted in Judeo-Christian traditions, colonialism, and narratives of escapism, romanticism, primitivism, and nationalism.[16] These culturally specific values reinforce a simplified perspective of wild places, causing ongoing social and political conflicts, such as debates on oil rights in Alaska, conflicts between the working class and environmentalists' efforts to protect "pristine nature," and long-term tensions between Indigenous people's homelands and North America's wilderness. Early settlers saw no organization and no signs of control because, in their eyes, land management implied clear signs of agriculture. However, evidence shows that Indigenous hunter-gatherer cultures had intensely managed their land, using a system other than agriculture. In a way, hunter-gatherers practice a form of "farming" that requires people to intentionally care for and manage the nonhuman species around them in order to harvest them.[17] After digging matsutake mushrooms, human pickers cover the ground to protect the mycelium networks beneath; in the next growing season, these networks will produce more matsutakes. Though discrete and sporadic in seemingly wild places, this is still a form of land management that shares an underlying logic with farming—to care for another species and hope to repeatedly harvest them.

Despite the intellectual criticism of binary thinking, contemporary environmental practices are still awash with dualisms, such as artificial/natural and technology/environment. They lead to moral justifications that idealize "nature-based" approaches and vilify "technological intervention," oversimplifying environmental strategies. If wildness entails a lack of human influence, how might intelligent machines, regarded as extensions of human capacities, construct wild places? A framework that strictly divides the technological and the biological makes it difficult to consider machines and wildness within the same category.

The boundary between technology and nature is often more apparent in our minds than in the environment itself. It is important to recognize that many wild places we refer to as "nature" require maintenance and care to exist. For instance, New York City's Central Park is often considered "nature in the city," but it is a highly constructed and maintained urban landscape. Moreover, contemporary posthumanist ideas, such as assemblage theory, actor-network theory, and new materialism, share a co-production thesis for understanding human agency. These theories propose that human agency and intelligence are distributed throughout a heterogeneous meshwork of assemblages, including humans, nonhuman species, and machines, as well as organizations, corporations, policies, and laws. Contemporary landscape design and urban planning theories also recognize that the urban environment is co-produced and shared by humans and nonhuman species; actors' actions modify each other, resulting in a mesh of different perspectives.[18] The outcome of co-production often lies beyond the actors' original intentions, if they existed at all. Therefore, *wildness* describes perceived wild conditions with unexpected outcomes that often refuse to fit within the agents' mental models.

Furthermore, we can explore wildness in relation to autopoiesis and second-order cybernetics. As Maturana and Varela argued:

> [w]e as observers have access both to the nervous system and to the structure of the environment. We can thus describe the behavior of an organism as though it arose from the operation of its nervous system with representations of the environment or *as an expression of some goal-oriented process*. These descriptions, however, do not reflect the operation of the nervous system itself. They are good only for the purpose of communication among ourselves as observers.[19]

When we adopt this highly recursive perspective, cybernetics becomes an epiphenomenon that observers use to describe and categorize system behaviors. The concept of goal-directness, in this view, is a construct invented by human observers to establish a cause-and-effect relationship between agents' actions and their consequences. In other words, it is not that cybernetic systems with goal-seeking behaviors exist independently of our observations and categorizations; rather, we invent and adopt cybernetics to explain the world in terms of agents and goals and then identify phenomena that fit within this descriptive category.

For instance, machine learning may need to be defined among scientists as an agent seeking goals, which can be articulated through reward functions. However, evaluating and measuring these functions are actions distinct from a neural network's system operations. While we may assert that agents have goals and utilize the imagined goal-directed behavior to implement control strategies, cybernetics and goal-seeking behaviors are insufficient to explain agent behaviors.

Drawing on OOO, we cannot access a real object that is always withdrawn from access, even with cybernetic thinking and systems theory. Agents' system operations modify, amplify, dampen, and overlap with each other, resulting in a sense of wildness in the co-produced environment where the outcome is alien to all.

Thus, wildness and unpredictability should be viewed as the new norm of the environment in which designers operate. From this new baseline, intelligent machines need not be regarded as a layer of a control mechanism through which humans extend imagined agency and expand an illusory control regime. Instead, intelligent machines may be conceived of as multi-scalar actors deeply embedded in environmental processes, contributing to the creation of a dynamic yet erratic environment. Thus, conceptually, the "environment" which today's designers must consider fundamentally differs from those on which past designers concentrated. "Environment" no longer stands for a passive background or a *tabula rasa* on which designers entertain system dynamics. Rather, designers confront a cybernetic environment embedded with diverse forms of distributed intelligence. Design should no longer use localized human intelligence, an illusion from a posthumanist perspective, to solve environmental problems. To design is to recognize the distributive nature of intelligence, acknowledging other forms of "mental processes" that differ from ours.

With this new framing, cybernetics no longer leads to controlled stability, but to a *cultivated wildness*; feedback mechanisms between agents lead to conditions unexpected and alien to all. This is unexplored territory where new questions arise, new understandings form, and new strategies emerge. Below is an illustration of a conceptual framework, inspired by machine-learning algorithms, to explore the notion of *cultivated wildness* (Figure 13.3).

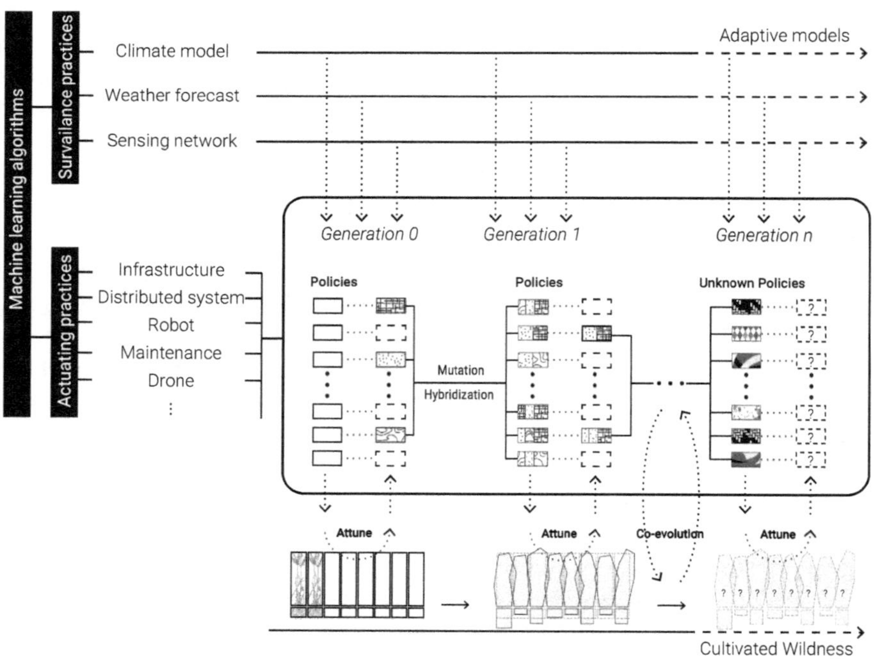

FIGURE 13.3 Cultivated Wildness: A Conceptual Roadmap. Illustration by the author.

Within this framework, machine-learning algorithms are integrated into cyberphysical systems. These include environmental surveillance practices like climate models, weather forecasting systems, and sensing networks, and actuating systems, such as cyberphysical urban infrastructures, robotic armatures, distributed robots, and drone swarms. These algorithms interact with nonhuman assemblages, including plants and animals, and may influence their habitats by modifying soil conditions and hydrological patterns. Machine-learning agents employ continual learning and genetic algorithms, as well as transfer learning, to test and improve their strategies, while being informed by surveillance practices and past experiences. These distributed and loosely coupled agents are capable of developing unexpected strategies over time. Landscape architects, engineers, and designers can contribute to this framework by developing interfaces and media to communicate with these multi-scalar models and systems. Within this framework, an intelligent machine is not a tool for human control, but rather a necessary medium through which we can explore the unrealized potential of the cybernetic environment.

Notes

1 Morton, "Ecology Without Nature: Attunement." https://ecologywithoutnature.blogspot.com/2014/11/attunement.html. Last accessed: April 29, 2024.
2 Estrada, "Towards Sentience: Attuning the Los Angeles River's Fluvial Morphology" https://www.gsd.harvard.edu/project/8531/. Last accessed: April 29, 2024
3 Estrada, "Prototyping." See https://towardssentience.com/prototyping. Last accessed: April 29, 2024.
4 Estrada, "Towards Sentience."
5 Gonzalez, "Design for a Mind with Many Bodies." See https://dspace.mit.edu/handle/1721.1/106370. Last accessed: April 29
6 https://wyss.harvard.edu/media-post/romu-a-robot-for-environmental-protection. Last accessed: April 29
7 Brown et al., "Language Models Are Few-Shot Learners."
8 https://help.openai.com/en/articles/8554397-creating-a-gpt. Last accessed: April 29
9 Chen, Murali, and Gupta, "Hardware Conditioned Policies for Multi-Robot Transfer Learning."
10 Barekatain, Yonetani, and Hamaya, "Multipolar."
11 Bijker, "How Is Technology Made?," 69.
12 Bowen, *Growth Rendering Device*. https://www.dwbowen.com/growth-rendering-device. Last accessed: April 29, 2024.
13 Bowen, *Growth Modeling Device*. https://www.dwbowen.com/growth-modeling-device. Last accessed: April 29, 2024.
14 Bowen, *Plant Drone*. https://www.dwbowen.com/plantdrone. Last accessed: April 29, 2024.
15 Cronon, "The Trouble with Wilderness" 72.
16 Cronon, "The Trouble with Wilderness"
17 Graeber and Wengrow, *The Dawn of Everything*.
18 Houston et al., "Make Kin, Not Cities! Multispecies Entanglements and 'Becoming-World' in Planning Theory"; Prominski, "Andscapes."
19 Maturana and Varela, *Autopoiesis and Cognition*, 129. Emphasis added.

Bibliography

Barekatain, Mohammadamin, Ryo Yonetani, and Masashi Hamaya. "Multipolar: Multi-Source Policy Aggregation for Transfer Reinforcement Learning between Diverse Environmental Dynamics." *arXiv:1909.13111 [Cs, Stat]*, September 28, 2019. http://arxiv.org/abs/1909.13111.

Bijker, Wiebe E. "How Is Technology Made?—That Is the Question!" *Cambridge Journal of Economics* 34, no. 1 (January 1, 2010): 63–76. https://doi.org/10.1093/cje/bep068.

Bowen, David. *Growth Modeling Device*. 2009. https://www.dwbowen.com/growth-modeling-device.

———. *Growth Rendering Device*. 2007. https://www.dwbowen.com/growth-rendering-device.

———. *Plant Drone*. 2019. https://www.dwbowen.com/plantdrone.

Brown, Tom B., Benjamin Mann, Nick Ryder, Melanie Subbiah, Jared Kaplan, Prafulla Dhariwal, Arvind Neelakantan, et al. "Language Models Are Few-Shot Learners." *arXiv:2005.14165 [Cs]*, July 22, 2020. http://arxiv.org/abs/2005.14165.

Chen, Tao, Adithyavairavan Murali, and Abhinav Gupta. "Hardware Conditioned Policies for Multi-Robot Transfer Learning." *arXiv:1811.09864 [Cs]*, January 12, 2019. http://arxiv.org/abs/1811.09864.

Cronon, William. "The Trouble with Wilderness: Or, Getting Back to the Wrong Nature." In *Uncommon Ground: Rethinking the Human Place in Nature*, edited by William Cronon, 1st ed., 69–90. New York; London: W. W. Norton & Company, 1996.

Estrada, Leif. "Prototyping." Towards Sentience™, 2016. https://towardssentience.com/prototyping.

———. "Towards Sentience." In *Codify: Parametric and Computational Design in Landscape Architecture*, edited by Bradley Cantrell and Adam Mekies, 279–88. Milton Park; Abingdon; Oxon; New York: Routledge, 2018.

———. "Towards Sentience: Attuning the Los Angeles River's Fluvial Morphology." Harvard Graduate School of Design, 2016. https://www.gsd.harvard.edu/project/8531/.

Gonzalez, Ricardo Jnani Ramirez. "Design for a Mind with Many Bodies : Cybernetic Micro-Interventions in the Cryosphere." Thesis, Massachusetts Institute of Technology, 2016. https://dspace.mit.edu/handle/1721.1/106370.

Graeber, David, and David Wengrow. *The Dawn of Everything: A New History of Humanity*, 1st ed. New York: Farrar, Straus and Giroux, 2021.

Houston, Donna, Jean Hillier, Diana MacCallum, Wendy Steele, and Jason Byrne. "Make Kin, Not Cities! Multispecies Entanglements and 'Becoming-World' in Planning Theory." *Planning Theory* 17, no. 2 (May 2018): 190–212. https://doi.org/10.1177/1473095216688042.

Maturana, H. R., and F. J. Varela. *Autopoiesis and Cognition: The Realization of the Living*. Boston Studies in the Philosophy and History of Science. Cham: Springer Netherlands, 1980. https://doi.org/10.1007/978-94-009-8947-4.

Morton, Timothy. "Ecology Without Nature: Attunement." *Ecology Without Nature* (blog), November 2, 2014. http://ecologywithoutnature.blogspot.com/2014/11/attunement.html.

Prominski, Martin. "Andscapes: Concepts of Nature and Culture for Landscape Architecture in the 'Anthropocene.'" *Journal of Landscape Architecture* 9, no. 1 (January 2, 2014): 6–19. https://doi.org/10.1080/18626033.2014.898819.

14

CULTIVATED WILDNESS AND SPECULATIVE ECOLOGY

Non-dualistic Thinking and Technodiversity

Cultivated wildness challenges the illusory boundary between the biological and the technological by overcoming dualistic thinking. Yuk Hui argues that cybernetics is thinking *par excellence*, providing a universal theory for dissolving the boundary between biology and technology. As a type of organicism, cybernetics critiques "mechanism" as the ontological framing for machines.[1] Cybernetics constructs a universal theory based on systems to describe recursive behavior across all entities, as suggested in Norbert Wiener's book *Cybernetics: Or Control and Communication in the Animal and the Machine*. This is further evidenced by the early cyberneticians' machines that mimicked animals and biological behaviors—William Grey Walter's cybernetic tortoise, Wiener's "moth and bedbug," and Von Neumann's cellular automaton. By the mid-twentieth century, cybernetics surpassed the mechanistic view, opening a range of exploration for modern machines as self-regulating organisms. Nearly all modern machines are based on a circular causality and feedback mechanism, which links them to biological entities that determine themselves through recursive structures.[2] This undertaking has carried on through systems theory, an irreplaceable epistemic substrate underpinning contemporary intellectual life.

The problem, according to Hui, is that cybernetic thinking attempts to dissolve the boundary between the biological and the technological by unifying them through a synthesized vision. Hui criticizes the dualistic thinking that arises from trying to reconcile the tension between the two realms. By unifying the tensions, we construct a field in which we are comfortable, a situation that does not threaten us. Yet at the same time we create the other side of

DOI: 10.4324/9781003320852-18

the field, which we perceive as threats that lie outside our experience and for which we have no vocabulary. In doing so, we create a new unexplored area of thinking that is neither cybernetic nor systemic.

From this perspective, cybernetics may have offered a framing in which the tension between the biological and the technological is no longer acute, under the overarching metaphor of "system." Yet we create "non-cybernetic" thinking, occupying the other side of our conceptual field, where our language falls short. As Hui notes, "Cybernetic thinking remains a thinking of totalization, since it aims to absorb the other into itself, like Hegelian [dialectical] logic, which sees polarity not as oppositional but rather as a motivation towards synthesized identity," and therefore, "to think beyond cybernetics is to think beyond the totalizing effect of a non-dualist thinking."[3]

We are far from overcoming dualistic thinking. This field featured by "non-cybernetics" becomes an antithesis that challenges us to resolve it with another level of synthesis. If cybernetics is about control and communication between actors, then the other side of the field is where things are out of control, do not communicate, and are withdrawn. It is the field of wildness—the spectrum between control and wildness results from the very dualistic thinking that compelled the development of cybernetics. From this vantage, "cultivated wildness" seeks to provide a mode of thinking that surpasses the totalizing effect of cybernetics. We may examine this effort from two levels.

First, it is crucial to recognize that controllability is a concept that seeks to reconcile the opposing forces of control and wildness by creating a spectrum of varying degrees of control. It still imposes a single vision to resolve the antithetical pair that occupies the two sides of our conceptual field. In contrast, the notion of "cultivated wildness" is an ontological claim that accepts control and wildness as a pair of polarities comprising reality itself. It is an ontology because it concerns the ability—or willingness—to accept the conflicting aspects of a situation for what they are. We can find wildness in control, and vice versa.

With this framing, we must then acknowledge that control and wildness are at all times in dynamic transition. This transition does not involve relativism or different percentages of control and wildness canceling each other. It accepts that when we control wildness, we simultaneously construct it. In other words, "cultivated wildness" reconfigures the notion of control and agency. We begin to see no difference between what it is to control and what it is to relinquish control; we perceive non-action as action and see a lack of agency as a sense of agency.

We can test this framing by analyzing present-day popular concepts, such as resilience. We may think of a flood wall as an ultimate control strategy for urban resilience.[4] Flood walls keep excess water out, but when they block its passage, that water must flow somewhere; this "somewhere," omitted from our field of controllability, becomes an intrinsic part of the control strategy itself. Certain landscape architects may find this mode of thinking familiar. Indeed, landscape architects work with living materials that possess a sense of wildness; they must conceptualize those outside their control horizon as part of their design strategies, and, in turn, view their strategies as a way to promote elements that are out of control, anticipating emergent behaviors. Thus, many landscape architects would agree that design does not mean doing everything; refraining from doing everything becomes a way of design. Unfortunately, this ambiguous and recursive reasoning fails to fit within the mainstream, modernist descriptive and interpretive frameworks of controllability that celebrate a designer's agency and abilities to act. Another large obstacle resides in Western individualism, which advocates for individual agency. More work must be done to develop vocabularies and concepts in order to open possibilities for non-dualistic thinking in contemporary environmental discourse. Thus, theorizing about landscape practices and reflecting on their meaning in contemporary culture is crucial for developing a unique epistemological framework for landscape architecture, among other disciplines.

Second, and on a different level, "cultivated wildness" exemplifies one way to critique an idea using non-dualist thinking, by embracing a sense of diversity and conflicting aspects within a unified framework. This book has re-conceptualized cybernetics by returning to its origins and offering new interpretations in light of contemporary posthumanism, specifically its ontological and epistemological concerns. Critiques of cybernetics may be situated within the broader intellectual reflection on relational and systems thinking present since the early 2000s. Cybernetic thinking is relational, based on the premise that entities communicate. Many contemporary philosophical ideas, such as object-oriented ontology (OOO), challenge this relational thinking by emphasizing the opposite side of the system—objects. This is why OOO emphasizes objects' non-communicative and withdrawn aspects, creating a framework based on the premise that things do not communicate. However, this type of criticism is still dualistic and attempts to construct an antithesis to replace relational thinking. "Cultivated wildness" does not claim to be a version of "non-cybernetic thinking." Instead, as we saw in previous chapters, it is strongly based on recursive causality and feedback mechanisms and, thus, cybernetic.

At its heart, the aim of cybernetics is to use recursive causality to frame mental processes across different entities. However, we must recognize that cybernetics has been interpreted in a societal episteme characterized by modernity. It entails society's unquestioning acceptance of technological progression and techno-scientific values, overlooking aspects other than how cybernetics could help build powerful machines. As a major descendant of the cybernetics movement in the mid-twentieth century, modern control theory presents itself as if cybernetics solely involves using feedback loops and communication behaviors to build powerful machines. It constructs a linear and deterministic view between control and communication, in the sense that communication leads to control, and control leads to stability. When doing so, we eliminate the possibility of exploring alternative relationships between control and communication.

There are precedents for reinterpreting cybernetics in a different context, outside mainstream concerns. In the 1970s, when introducing the "scoring system" and RSVP design methodology, landscape architect Lawrence Halprin posited that scores "communicate but do not control."[5] This was a clear riposte to Wiener's *Control and Communication in the Machine and the Animal*. Halprin's framing is advanced even for today's readers. Consider this passage:

> One of the gravest dangers that we experience is the danger of becoming goal-oriented… it is a trap which goes like this: things are going poorly… As thinking people we must try to solve this problem that faces us. Let us set ourselves a 'goal' upon which we can all agree (most goals after all are quite clearly moralistically based and incontrovertibly 'good ideas'). Having set ourselves this goal we can then proceed posthaste to achieve it by the *most direct method possible*. Everyone can put his[sic] shoulder to the wheel, and systems engineering, technology, and our leader will get us to the agreed goal… It doesn't work![6]

This passage can be viewed as a critique of the goal-directedness intrinsic to cybernetics.

From a contemporary perspective, Halprin attempted to develop a theory akin to second-order cybernetics, autopoiesis theory, and emergence. However, second-order cybernetics and autopoiesis theory were in their infancy in the early 1970s, and the notion of emergence did not gain currency until the 1990s. In a discipline that might be described as a modern other,[7] Halprin's concern was overwhelmed by ideas that better fit within the descriptive and interpretive categories of mainstream cybernetics characterized by controlled stability.

In contrast to Halprin, Ian McHarg was an important figure in the mid-twentieth century, who brought cybernetic thinking into the landscape discipline via ecological science.[8] McHarg's science-inspired design frames a more deterministic and goal-oriented view of landscape dynamics. Ecosystems exhibit certain cybernetic qualities and will always experience ecological successions and reach climax communities, which are considered stable and ecologically fit. This framing yields the most direct method possible to achieve the goal of saving the environment. McHarg's homeostatic interpretation echoed the mainstream expectation for controlled stability produced by cybernetic machines. Thus, the impact and acceptance of McHargian design methodology were undeniably more profound than those of Halprin, particularly in a global context.

From this vantage, this book has furthered Halprin's legacy and provided a disciplinary critique and reflection on cybernetics, with the privilege of posthumanist vocabularies and concepts. However, we must recognize that "cultivated wildness" is not the ultimate cybernetics, but a version of unexplored cybernetics. It exemplifies a viable method of interpretation and merits further investigation in order to discover versions of cybernetics thinking that may conflict with and simultaneously complete each other. This leads to the discussion on technodiversity.

We have briefly introduced "ecology of machines" and "technodiversity," proposed by Hui, as two important concepts with which to examine the cases presented in Chapter 11. We shall further develop these two concepts and investigate their implications to an adaptive epistemology. As Hui noted, the foundation of ecosystem ecology is a sense of biodiversity, which sets a basis for multi-species interaction. In parallel to biodiversity, Hui proposes technodiversity as a foundation for the ecology of machines.[9] It is not concerned with different ways to use a machine, and the "ecology of machines" is not about different types of machines interacting with each other. Instead, these two concepts entail a diversity of technical frameworks, or different ways to approach a technical framework. "Cultivated wildness" thus represents a sense of technodiversity, as it seeks to provide an alternative way to approach cybernetics in light of posthumanism.

Let us extend the analogy between biodiversity and technodiversity. The mainstream interpretation of cybernetics has itself deployed a monocultural approach to thinking—one that is based on controlled stability—eliminating other possibilities for interpretation. Consequently, we have cultivated a monoculture of machines conceptualized as the means to automate and replicate human physical and mental capacities in order to control. Thus, a convenient way to consider machines in relation to ecology is to regard them as powerful tools to increase biodiversity and recover wild places. However, this falls into an ecomodernist trap, which reinforces an outdated transformation formula in which, using technology, humans transform pristine nature into habitable landscapes. However, as argued, we have co-evolved with tools and machines, through which we construct relationships with other species. Thus, to consider cybernetics along with ecological thinking is not about promoting biodiversity with intelligent machines, but cultivating technodiversity and embracing a sense of "wildness in machines."

Extending this metaphor, preservation and conservation efforts should do more than preserve "wild nature"; they must also preserve techniques through which we engage with the environment. From this perspective, biodiversity relies heavily on technodiversity. In recent years, research in indigenous studies has presented vivid evidence of different ways in which a society can relate to its environment other than those developed during colonialism.[10] Thus, cultivated wildness merits further exploration in its relationship with decolonization, decoloniality and pre-colonization. The way we represent and construct the environment is inseparable from the way we choose to live in it, including our modes of thinking that choose the technical frameworks we use to build machines. Therefore, "wildness in machines" challenges us to rethink technological development and to investigate the unexplored realm, even in a mundane machine, speculating on its unrealized potential.

Eventually, "wildness in machines" requires a sense of "speculative ecology" toward the notion of diversity. Diversity is not about seeing difference as a motivation toward a synthesized framework; instead, it is about continual and life-long efforts to live with and attune ourselves to the different frameworks with which we have no experience. Diversity speculates about different ecologies among beings. Thus, landscape architecture provides a venue for a speculative ecology.

Landscape Architecture as Speculative Ecology

Challenges for a Landscape Epistemology

Previous chapters have embraced a liberal understanding of "design" in order to identify the shared cybernetic thinking underlying art, landscape design, and engineering. This conceptual move may be essential for an alternative framing of cybernetics, but it dilutes the term "design" and ignores the nuances with which different fields approach "design" along with their specific disciplinary concerns. Now is the time to emphasize the unique and transformative thinking which contemporary landscape architecture might offer to design disciplines.

It must be recognized that landscape architecture has struggled with its identity within the modern disciplinary tradition. This struggle emerges at two levels. The first concerns the general identity crisis of "design" professions. Designers distinguish between those who actually "practice design" and those who regard "design" as a subject of inquiry. Over the years, concepts such as design thinking have gained currency across disciplines. The popularization of design can be understood as a process to "scientise" it.[11] This process embodies society's aspirations to apply techno-scientific frameworks in understanding the creative operations found among "designers" such as architects and engineers. The 1960s was known as "the design decade," following architect and technologist Buckminster Fuller's call for a "design science revolution." The 1962 Conference on Design Methods, held in London, may be regarded as the starting point of design methodology as a field of inquiry. However, this view was challenged from the beginning. Certain pioneers in the 1960s design methodology movement have even distanced themselves from this view.[12] The irony is that designers, notably architects and landscape architects, seldom practice a diluted, structured, and linearized "design thinking" process.

Despite the backlash, design methodology research continued to develop over the following decades; today, design thinking is a popular concept freely discussed by everyone but designers. This has given rise to a sort of identity crisis for contemporary designers: while they wish to be identified as designers, they avoid being stereotyped as "designers" constructed in the field of design methodology. Nigel Cross uses the term "designerly ways of knowing and thinking" to encapsulate the urge to identify design as separate from other forms of knowledge production. "Design" may become a discipline unto itself, with unique epistemic frameworks and concerns.[13] Many would agree that design, as a way of knowing, deserves its own vocabularies, concepts, and disciplinary frameworks. However, the struggle we face is the need for locality and diversity within the urge to reach a synthesized term for the alleged "design epistemology." By using the term "design," we must recognize that we refer to an extremely diverse community that contains a plethora of approaches to this non-scientific, non-engineering, non-artistic way of knowing and thinking. The issues of locality and diversity lie at the heart of today's "design research" discourse.

The second struggle arises from the marginalized position of landscape architecture among modern disciplines. As landscape theorist Elizabeth Meyer has argued,

> As a field that built physical critiques of, and in, the American city that embodied broader society's unquestioning acceptance of industrialization and technological progress, landscape architecture has not fit within the descriptive, evaluative, and interpretive categories of mainstream modernism—historical or theoretical.[14]

This marginalized position forces landscape architecture to turn to models from other fields to articulate its internal complexity in theories and methods. Contemporary landscape practices in North America largely operate based on two models developed since the mid-twentieth century: modern art/architecture and ecological science.[15] Even though, since the late 1990s, many landscape architects have already bridged the gap between the two, expanding the field of landscape architecture, the concepts and frameworks of these two models are preeminent. For example, Ian McHarg's ecological design framework bears a strong legacy through ideas such as "landscape urbanism."[16]

Of course, contemporary landscape architecture has bypassed the ecological concerns of McHarg's time, those that revolved around homeostasis and determinism. Instead, landscape architecture today embraces a nonlinear and non-deterministic view focused on emergence and novel ecologies.[17] However, these new conceptualizations rely on ideas and conclusions from the modern ecological sciences. To a certain extent, today's landscape architecture persists in borrowing thinking from exterior models, without forming an epistemic framework unique to the field.

Yet another factor contributing to the identity crisis comes from the fact that landscape thinking itself has been transformed into a model by modern design professions, and "rebranded" by adjacent fields, including architecture and urban design, into a "new" framework for urbanization (e.g., "landscape urbanism"). With its focus on infrastructure and urban ecology, landscape urbanism has become a dominant doctrine in contemporary landscape practices. It is still essentially a McHargian ecological determinism merged with economic determinism.[18] The problem is that these doctrines reduce landscape architecture without recognizing its intrinsic complexity. The simplified models, in turn, become landscape architects' self-scrutinizing and self-sanctioning mechanisms, which legitimize, and favor certain bodies of work and limit the possibilities of other epistemic frameworks unique to the discipline.

A Nonhuman Turn and Posthumanist Landscape Epistemology

Recent years have witnessed a thriving interest among the design disciplines, including landscape architecture, in searching for new vocabularies and frames with which to articulate "designerly" epistemic concerns, distinct from those of the arts, sciences, and engineering. In landscape architecture, this interest manifests as a "nonhuman turn" embodied in skepticism about designers' agency and control, and an urge to include and embrace diverse frameworks, especially those provided by nonhumans, in shaping the landscapes. This may be described as an advance toward a posthumanist landscape framework.

Many landscape scholars have begun to deploy various theories, such as Donna Haraway's "companion species," to conceptualize animals and plants as active agents shaping landscapes, rather than as visitors or inhabitants providing ecosystem services.[19] This trend can be situated within a broader intellectual movement found in philosophy, environmental humanities, and science and technology studies (STS) since the 1990s. This burgeoning movement revolves around post-structuralist and posthumanist thinkers, including Bruno Latour, Donna Haraway, Jane Bennett, Cary Wolfe, Manuel DeLanda, Graham Harman, and Timothy Morton. As landscape scholars turn their attention to this trend, many of these thinkers have become widely read and listened to by both practitioners and theorists. For example, over the years, Harman and Morton have lectured at many North American design schools. Both have taken on teaching positions within SCI-Arch's architecture and landscape programs. Morton also served as a jury member for a prominent landscape competition in 2020, and his writings and thoughts on ecology have impacted the landscape discipline.

The core of this "nonhuman turn" in landscape architecture challenges designers to re-conceptualize modes of knowing and doing. Mainstream design methodology research often contrasts design with science, regarding science as a process of finding patterns in the universe, and design as constructing new patterns.[20] At the heart of this understanding is a division of theory and practice: the former belongs to the realm of abstraction, the latter to application. As Nigel Cross argues, design research has shifted from the urge to create a "design science" to creating a "design discipline" with unique "designerly" ways of knowing and thinking. This urge manifested in the landscape discipline as an emphasis on its unique way of knowing, with particular consideration of the nonhuman realm.

In recent years, the most transformative works have emerged from Brian Davis's discussion on landscape instrumentalism, and Cantrell and Holzman's consideration of "modification."

Drawing ideas from John Dewey's instrumental philosophy, Harman's OOO, and Bennett's discussion of vibrant materialism, Davis uses the term *instrument* in a manner akin to the idea of an actor or intelligent agent. He refuses to "reduce everything to a mere tool that does only what something else intends."[21] Consider these paragraphs:

> a landscape is made of instruments [agents] whose actions never align perfectly with a user's intention but are always doing more and less, creating a liminal space between intent and reality.[22]
>
> ...if a space is a landscape, and not some other type of space, then all of its objects and their dynamic relations are instruments, but not dumb drills, retaining walls, and land use policies. Rather, they are dynamic objects in relation to one another within a bounded territory containing some measure of human intent.[23]

Cantrell and Holzman gave a similar account when conceptualizing *modify* as a strategy in the Responsive Landscapes framework, suggesting a mode of inquiry between intent and reality, as here,

> ... modify is not to translate an existing condition or to construct condition—rather, through direct and indirect behaviors adjustment, it suggests *recursive inquiry through repetitive action.*[24]

Both "instrumentalism" and "modification" search for a sense of the posthumanist framework unique to landscape discipline. However, they share a drawback that undercuts the potential for their consideration to be transformative. Both of them can be described as "humanist posthumanism" because, though we seek to look beyond our limited human understanding of the environment, our thinking still relies on all-too-human terms to conceptualize this posthumanist framework. To a certain extent, this predicament results from the "philosophy of access," specifically, the philosophy of human access that undergirds many contemporary posthumanist considerations, including vital materialism, ANT, and assemblage thinking. In other words, the epistemic undergirding of our current consideration still privileges the capacity of (human) knowing as a basis upon which to conceptualize landscape design, even though we wish to subscribe to nonhumans for help. This drawback may be further situated within a broader philosophical context, or "the prevalent tendency with Kantian and post-Kantian thought to treat the *relation* between thought and world as the primary subject matter of philosophy."[25]

There are two levels of reflection. The first is about *human* access. Both "instrumentalism" and "modification" suggest a strong motivation for developing new techniques and methodologies to probe the liminal space between intent and reality, to inquire about it. Following the philosophy of human access is a type of means-end reasoning in considering landscape inquiries concerning tools and techniques. This may be observed from Cantrell's and Holzman's assertion that responsive landscapes provide frameworks that *utilize* responsive technologies as a form of *inquiry,* because "the contemporary landscape, as a site of emergent and novel conditions, undoubtedly holds conditions that have escaped previous attempts of translations."[26] Davis develops Dewey's instrumental theory of knowledge as a "theoretical and technical framework to develop the tools needed to integrate and synthesize new techniques for modeling, computation, analysis and construction of multi-functional landscapes, and to chart a way forward into the frontier between intent and reality."[27] A strong sense of means-end reasoning exists in both cases; tools and techniques, including intelligent machines, are considered effective means for designers to make inquiries. Designers may "instrumentalize" different actors in order to modify and study the environment through repetitive actions. Following this reasoning, cybernetic technologies are the means through which we "know" the world, exercising the agency of inquiry. To some extent, these considerations still insist upon a modernist view that privileges the concept of progress; to know more also means to expand a human-centered epistemology into nonhuman realms. This reasoning creates a paradox that undercuts the moral incentive to develop a posthumanist landscape framework. By including nonhuman species within our instruments of inquiry, we promote and justify yet another level of human hubris advanced by our alleged extraordinary ability to know.

The posthumanist undertaking aims to cultivate a sense of humility in designers by reflecting on their agency in control, encouraging designers to consider and embrace those frameworks unaligned with ours. On this level, both "instrumentalism" and "modification" have, without doubt, provided exceptional critiques and reflections. Regrettably, both of their responses to the initial insights fall short, by clinging to the last hope for "human access and knowing."

The second level of reflection concerns the philosophy of access—not only human access but also all sorts of access between objects. Popular concepts, such as ANT, assemblage thinking, and new materialism, which undergird contemporary concerns in the landscape discipline, are all relational. They are all based on the premise that, through interaction, objects are able to communicate and form relational bonds, enabling each other to gain more power (again, based on a progressive narrative). Relationships between them become *a priori* for elements to be regarded as significant. This is best exemplified by Jane Bennett's claim that "bodies enhance their power in or as a heterogeneous assemblage."[28]

Indeed, relational thinking comfortably aligns with the landscape discipline's ecological concerns, which are often understood as relational, active, and vibrant. A positive ontology is undoubtedly favored by the mainstream ideology of ecology and diversity, which encourages communication and interaction between individuals. The access philosophy reflects a sense of idealistic and occasionally romantic thinking about ecology and relationship. An underlying premise for "instrumentalism" and "modification" is that, with the right "instruments," technological or biological, landscape designers may cultivate a version of "ecology" and "interconnectedness" between actors, and this version of the web of assemblages will benefit all entities. This is an implied goal in the seemingly open-ended framework based on recursive inquiry and repetitive actions. Hence, Cantrell and Holzman ask, "[a]t what point are goals, scaffolds, and protocols actually open-ended?"[29]

Instead of attempting a solution to this question, we should reflect on why it became a question in the first place. The question suggests a mental model that creates a version of an interconnected network between human and nonhuman entities; in this version of ecology, humans are "in harmony" with other humans and nonhumans, all of which happily share an environment, "co-evolving" and "co-existing." This model is vivid enough to serve as a goal to motivate all sorts of techniques, methodologies, and frameworks to achieve this end. Ironically, this version of "co-existing" and "co-evolving" exemplifies another level of human exceptionalism, celebrating our supposed ability to *know* what other species seek, and *design* an appropriate ecology for them.

Eventually, this line of reflection leads to the notion of utopian thinking. To conceptualize design as constructing new patterns and rearranging existing structures sets a goal, a product, or a result at the end of a collection of creative protocols which we call "design." This utopian thinking also relies on a communicative framework, imagining that we can reach a better place through repetitive actions, inquiries, and communications. This type of idealistic thinking is fundamental to conceptualizing intentionality in design practices. As we saw, both "landscape instrumentalism" and "modification" rely on the notions of intentionality, conceptualizing an operational space that is liminal, between the intentional and the unintentional. It embodies a sort of dualistic thinking, viewing intentionality and unintentionality as mutually exclusive. The gap between intention and reality leaves space for entertaining the idea of increasing perceived human agency in terms of knowing and controlling, to realize an "intended outcome." Intention relies on desires, and fixes goals, interpreting unexpected outcomes as motivation for inquiring about "truth."

Toward a Non-access Philosophy

The philosophy of access and the relational episteme risk overlooking a spectrum of thinking based on the assumption that things cannot communicate and relate to each other in the first place, and that we cannot find truth through techniques of "inquiry." This expands Niklas Luhmann's (in)-famous assertion that "[h]umans cannot communicate; not even their brains can communicate; not even their conscious minds can communicate. Only communications can communicate."[30]

Luhmann's provocative statement is based on second-order cybernetics and autopoiesis theory, which posit that signals from the environment do not produce a difference, but merely trigger a series of responses to generate a reality within a system through system operations.[31] From this perspective, *communication* becomes an epiphenomenon of system operations, a concept that we humans, who are cybernetic organisms, invent to describe perceived cause and effect when interacting with other systems, and when systems interact with each other. Multiple realities of the environment are constructed by different cybernetic systems. When systems interact, they must shape their mental processes into communicative operations shared by different systems. In the end, systems do not communicate because communication occurs only in the realm of communicative operations.

We may further develop Luhmann's assertion with a touch of OOO. As Morton argues,

> Things exist in a profoundly 'withdrawn' way: they cannot be splayed open and totally grasped by anything whatsoever, including themselves. You can't know a thing fully by thinking it or by eating it or by measuring it or by painting it…[32]

Things being non-communicative is the starting point for OOO; it seeks to create a non-relational ontology to differentiate itself from other posthumanist thinking. Since objects are withdrawn, they are inaccessible by any means, including human inquiry and thinking. Then, any form of inquiry becomes speculation, and thus OOO is regarded as a proponent of the speculative realism movement in philosophy.

An object comprises more than the relationships it forms with other objects. The term "agency" becomes irrelevant in OOO's framework, because to imagine we can relate to each other provides the initial urge and motivation to further act on that relationship and control. To imagine that our thought (or anything's thought) can form a relationship with other objects is fundamental to conceptualizing agency. This form of reasoning has manifested itself in cybernetics and modern control theory; both are based on the premise that recursive processes lead to communication, and communication leads to control. However, when we embrace second-order cybernetics, autopoiesis theory, and OOO's perspective, we can frame communication and control differently, and re-conceptualize cybernetic thinking in a speculative ontology.

From this vantage, we must recognize that the "nonhuman turn" in landscape discourse has successfully challenged designers' agency and control by introducing nonhuman frameworks into our tradition of human design and inquiry. However, the "nonhuman turn" does not chart what a posthumanist paradigm looks like for landscape architecture. Thus, cultivated wildness operates within a speculative ontology which the nonhuman turn fails to achieve.

A Logic of Coexistence: Expectation and Intentionality

Many, including Jane Bennett, duly distance themselves from OOO's claims for its seemingly pessimistic ontology. However, the opponents of OOO miss the ease and joy it brings. Supported by philosophies of access, we are challenged to inquire. We thus hold expectations for other beings around us, hoping to communicate and thus control. Yet these expectations may also provide a source of anxiety if plans of action are not discharged as expected. Failure motivates further inquiries and we are challenged to march on a journey without hope.

However, the non-communicative assertion in cultivated wildness presents a completely different line of reasoning; it opens a speculative ontology for the design professions. At the center of this undertaking sits one simple realization: *We cannot communicate with nor control each other, but that is acceptable.* Since things are non-communicative, as Morton puts it, "the way things affect one another (causality) cannot be direct (mechanical), but rather indirect or vicarious: causality is aesthetic."[33] If we accept the non-communicative nature of beings as a default, then the flickering moments where we believe we *are* communicating become a type of aesthetic experience. This is why Morton argues, "OOO believes that reality is *mysterious* and *magical*, because beings are withdrawn and because beings influence each other aesthetically…at a distance."[34] Thus, since we never reach an ultimate truth about beings, we are free to speculate without fear of failure or anxiety regarding unfulfillable expectations. Philosophies of access are limited by their "inability—or better, unwillingness—to create a *speculative ontology* which moves beyond the narrow confines of what is given to our all-too-human modes of understanding."[35]

With this passive yet speculative ontology, design requires great humility when designers operate in a non-communicative and uncontrollable world. But humility brings a sense of comfort. Since beings have no obligations to communicate, it becomes our responsibility as designers to strive to align ourselves with and attune to other beings. We must understand and respect that other beings—biotic or abiotic—will possess inaccessible aspects, and that, therefore, they may always behave in unexpected ways. Imagine that we viewed the environment in this way. We might then begin to derive ease and joy from this non-communicative perspective because intentionality, a human-centric concept, becomes unimportant. We are relieved of our anxieties and worries that things might evolve in unexpected ways, because we accept from the start that we have no control over them, and, therefore, no expectations.

As we have seen, contemporary landscape discourse struggles in the space between intent and reality by perceiving *intentionality* and *unintentionally* as mutually exclusive. Now, with this non-communicative framework, we may offer a non-dualist way to see intentionality and unintentionality—not as mutually exclusive, but as a dynamic transition.

Let us consider the condition of a "cultivated wildness." Any intentional decisions are likely the source of unintended consequences and unexpected outcomes. Even the portion we consider an intended outcome is, in fact, a flickering moment when withdrawn objects appear to synchronize. However, these fleeting moments are ephemeral, because things have no obligation to keep attuned to us; indeed, they may rapidly move out of synchronization. This is when we observe decay and entropy. We must then make efforts to maintain these living, dynamic relations through *design*. This maintenance of living dynamics is not about control, because control entails our expectations that, because things are communicative, we can send control signals to persuade them to act as we intend.

To summarize, a logic of coexistence with nonhuman objects becomes simple: things do not have an obligation to communicate in the first place, and it is we stubborn human beings who wish to form meaningful ties with each other and with the nonhumans around us, primarily because we find efficacy and a sense of agency in these ties. We perceive homeostasis and a sense of stability in the relations we observe with other beings. It is then our obligation to attune ourselves to them, rather than communicate with and control others to maintain these ties on our behalf. Adaptation requires great humility in ourselves; the non-communicative nature of beings urges us to make efforts and stand ready to do whatever is needed to maintain relations with them, even abandoning our previous beliefs. This is another way to understand what it means to form an adaptive epistemology for landscape design. Adaptive epistemology is not about continuous learning through recursive inquiries but continuous unlearning that which we once believed true. From this vantage, we may reflect on the notion of co-production of the environment: finding ourselves in a situation where our ways of doing synchronize with others, we humbly and carefully make efforts to keep it so.

Landscape Expertise: Scales of Speculative Ecology

The search for a "designerly" way of knowing and doing is inherently a critique of scientism. This search may be situated within the broader postmodern skepticism championed through the late twentieth century, exemplified by the ascendance of the social construction of scientific knowledge and feminist science technologies studies. However, "situated and embodied knowledge" justifies ignorance and arrogance toward expertise. Anyone can promote opinions about anything; all opinions must be equally "truths" and "realities." Over the past several years, debates around climate change and the COVID-19 pandemic have been vivid examples of this post-truth crisis.

"We cannot live by skepticism alone," says David Collins.[36] The field of STS has seen another turn since the late aughts of the twenty-first century, and STS schools have begun to pay attention to expertise as "tools for an initial weighting of opinion."[37] Many useful ideas have been developed in this discourse, including "trading zones" and "interactional expertise," emphasizing interdisciplinary knowledge and hybrid expertise that were overlooked by the conventional definition of knowledge production.[38] Similarly, OOO proponent Graham Harman also turns to the notion of expertise. In a way, OOO's claim that things are withdrawn and cannot be accessed seems to be a critique of scientific inquiry. In fact, OOO respects scientific expertise. Though things cannot be studied fully, even by scientific techniques, some of us still perform better in certain areas than others. That is because they have spent enough time with their allied objects, and developed living, dynamic relations with them. Their lived experiences allow them to respond to different situations with confidence.

From this perspective, there is no difference between design and scientific research; they are both expertise in developing and maintaining dynamic and living relations with other beings and using lived experiences to respond to new situations. There is a sense of aesthetic experience when machine-learning scientists find a new architecture for building artificial neural networks.

To consider a "designerly" way of thinking is to claim design as a type of expertise, as in the sciences.

The expertise of landscape architecture can be summarized as a speculative ecology. The issue of scale is intrinsic to landscape architecture, due to the diverse projects in which contemporary landscape architects find themselves. Landscape architects have expanded the profession's consideration in a spatial dimension, from gardens to urban parks to territorial landscapes. Meanwhile, with the notions of ecology and evolution, time becomes another scalable dimension which landscape architects entertain—from geological time, such as the Anthropocene, to phenomena as ephemeral as desert fog.

With the rise of systems theory in landscape discipline, specifically second-order systems theory, another dimension of scales was articulated. Anita Berrizbeitia introduced "scales of undecidability" as a conceptual framework with which to analyze Toronto's Downsview Park design competition, which featured projects with flexible frameworks and open-ended strategies. "Scales of undecidability" refers to "a landscape's capacity for precision of form notwithstanding flexibility of program," or, more precisely, "a landscape's capacity to engage multiple systems of signification at different scales."[39]

When we consider the recent "nonhuman turn" and posthumanist movement in the landscape discipline, I propose to articulate landscape scales in another dimension—scales of speculative ecology. Landscape expertise enables designers to speculate on the assemblages involved in the co-production of the environment. To a certain extent, "scales of speculative ecology" is a continuation and development of contemporary landscape concerns about open-endedness and adaptivity.

To illustrate the notion of scales of speculative ecology, we may contrast it with scales of undecidability. Berrizbeitia's undecidability relies on constructing a space between intent and reality, assuming different degrees of designers' control authority between the two. Consider the following passage:

> ...we can conceptualize landscapes where there is space and time for process to unfold and for stable meanings to come forth. We can imagine landscapes with the 'right mixture of rigid structures, supple structures and self-organizing processes' that are in some ways the ultimate solicitation of chance, in others the ultimate suspension of process, and yet in others something in between.[40]

Berrizbeitia suggests that designers possess different degrees of control authority over the landscapes they design. However, she continues to advocate for something else by ending her consideration with this paragraph,

> Engaging scales of undecidability is a social and political strategy, a tool for interference against any proposed solution that would be permanently

> static and definitive on the site. Only then will the urban park become a reserve of possibilities to come, for futures we cannot now imagine.[41]

Berrizbeitia asks for more than degrees of control; rather, her appeal is for a speculative framework that surpasses designers' intentionality. Her consideration is somewhat limited by the underlying dualistic thinking that contrasts a designer's intentionality with unintentionality. This undertaking relies on access philosophies, with the premise that we can know and communicate with things to carry out our intent, and leave open-ended whatever is uncontrollable. This reasoning provides a false sense of hope as motivation for perfecting one's design intentions, leaving space to consider control strategies to maintain a perceived status quo rather than a true, open-ended, and adaptive landscape epistemology.

In contrast, "scales of speculative ecology" is rooted in a non-access philosophy and a speculative ontology, emphasizing the conviction that things do not communicate and thus we have no control over them. Any intimations of control and stability are flickering moments when things accidentally align and synchronize.

Then, we may reflect on the notions of ecology and ecological research. Ecological thinking is relational thinking; at its heart is a perception of interconnections between things. How, then, do we understand ecology within a non-communicative and non-access framework from which things are withdrawn?

One way is to frame ecological research as an expertise of speculation. Any ecological model we use to describe relations between things can never provide truth, but is itself a speculative design. Thus, ecology is intrinsically speculative, and landscape architects become those able to test these speculations with built forms and materials. From this perspective, concepts such as "ecological restoration" lack validity, because where we "restore" a landscape based on ecological principles, we are, in fact, testing a speculative design of relations which we wish to see.

"Scales of speculative ecology" demand two levels of consideration. The first is the extent to which designers may speculate on the different assemblages involved in the co-production of the environment. These assemblages include different social groups, nonhuman species, intelligent machines, land-use policies, and other living and non-living entities in the process of individuation. Assemblage here is akin to Davis's notion of "landscape instruments." Yet "instruments" point to tools of utility and inquiry, which leaves a great deal of space in which to consider human agency in knowing and accessibility.

As we saw in the artwork by artists such as David Bowen and Sougwen Chung, nearly all projects involve unusual juxtapositions of assemblages and objects. Speculative ecology thus supplies the willingness to speculate about connections between objects, across scales. Speculative ecology is the capacity to imagine that American consumers' preference for almond products might impact the wild bee population and agricultural landscape of central California. It is the capacity to connect RL agents and machine-learning techniques with emergent ecologies. It is also the capacity to connect community gardens to tobacco plants, ozone levels, and the hyper-object of climate change.

The second level of consideration is the complexity of relations between assemblages, about which designers may speculate. Though things cannot communicate and have no obligation to communicate, we may speculate on scenarios where their operations are temporarily synchronized. The ability to speculate on new connections between objects becomes an act of design, which leads to unrealized dynamic relations between things. The only way to test these assumptions is to construct connections between assemblages, to observe what sorts of real-life relationships they form. From this perspective, a built landscape is no longer the product of a design process but a grand ongoing experiment in conjecture.

From this vantage point, we may reframe the role of built landscapes in our relationship with the nonhuman realm, and rethink the notion of failure. As a design product, a landscape entails designers' idealistic intentions as to how things ought to be and how they ought to relate to each other. If events and assemblages do not appear or end as expected, the design fails. However, suppose instead that a landscape is understood as an experiment where we test different arrangements of assemblages and observe how they interact—or fail to interact—with each other. In that case, there is no failure, but only unintended interactions generated from intentional design. These unintended interactions may become areas of potential. It may be that things synchronize and form meaningful relations, but we must be prepared to see otherwise, because we must respect that beings do not necessarily communicate with each other, and they will behave unexpectedly. Meaningful relations may be liminal and transient, and we must be resourceful enough to respond to a new situation.

How, then, do we cultivate the skills to be resourceful in any given situation and engage the scales of speculative ecology? To rephrase, how do we cultivate landscape expertise? I argue that there are two categories of techniques: *prototyping research* and *theorizing practices*. Both techniques require situating one assemblage or object within different frameworks, and exploring its unrealized relations.

Speculative ecology suggests that design requires constant engagement through which designers attune themselves to other beings. This requires a reworking of the mainstream design workflow that perceives a designed landscape as a product. Instead, designers must focus on maintenance as a preeminent strategy. Perhaps landscape designers should return to one of the discipline's traditions, and reflect on the importance of *landscaping* and *gardening* in contemporary landscape theory and practice. Perhaps the discipline must explore theories that support these ideas. Commonly used strategies such as phasing and maintenance offer an initial point from which to envision a landscape as a place where constant engagement occurs. Here, maintenance entails more than human and machine labor invested in the landscapes; it includes land-use policies, funding, organization, and the agencies behind a landscape project. In this way, we may begin constructing narratives of a landscape as a territory where strategies, protocols, and goals are truly open-ended.

This workflow will require the development of new tools and techniques with which designers may continuously code, choreograph, and attune to objects that compose the cybernetic environment. Environmental sensing, machine learning, and cyberphysical infrastructures may be incorporated into "landscaping" practices. Thus, the proposed theoretical framework may be greatly enhanced by prototyping research with machines. The key is to bypass the paradigm that sees machines as a layer of infrastructure to regulate and control "environmental systems." Instead, we must understand machines as agents deeply embedded in the landscape and interacting with other human and nonhuman beings. Prototyping is also concerned with exploring the "wildness" in machines and embracing a sense of "machine perspective"—the unexpected outcomes that lie outside designers' goals and challenge their conceptions of the cybernetic environment.

Notes

1 Hui, "Machine and Ecology."
2 Hui.
3 Hui, 63.
4 Even though many believe and promote the prioritization of "nature-based strategies" for resiliency, the irony is that prioritization means that "hard infrastructure" is the last resource when "nature-based strategies" fail, or that the place we wish to protect is too important to risk using "nature-based" solutions.
5 Halprin, *The RSVP Cycles*, 19.
6 Halprin, 4.
7 Meyer, "Situating Modern Landscape Architecture (1992)."
8 Lystra, "McHarg's Entropy, Halprin's Chance."
9 Hui, "Machine and Ecology."
10 Graeber and Wengrow, *The Dawn of Everything*.
11 Cross, "Designerly Ways of Knowing."
12 Cross.
13 Cross.
14 Meyer, "The Expanded Field of Landscape Architecture," 70.
15 Meyer, "The Post–Earth Day Conundrum: Translating Environmental Values into Landscape Design."
16 Steiner, "The Ghost of Ian McHarg"; Spirn, "Ian McHarg, Landscape Architecture, and Environmentalism: Ideas and Methods in Context."
17 Reed and Lister, *Projective Ecologies*.

18 Steiner, "The Ghost of Ian McHarg."
19 Klosterwill, "The Shifting Position of Animals in Landscape Theory."
20 Cross, "Designerly Ways of Knowing."
21 Davis, "Landscapes and Instruments," 294.
22 Davis, 294.
23 Davis, 305.
24 Cantrell and Holzman, *Responsive Landscapes*, 254. Emphasize added
25 Young, "On Correlationism and the Philosophy of (Human) Access," 43.
26 Cantrell and Holzman, *Responsive Landscapes*, 253.
27 Davis, "Landscapes and Instruments," 305.
28 Bennett, *Vibrant Matter: A Political Ecology of Things*, 23.
29 Cantrell and Holzman, *Responsive Landscapes*, 254.
30 Luhmann, "How Can the Mind Participate in Communication?"
31 Maturana and Varela, *Autopoiesis and Cognition.*
32 Dark Ecology: For a Logic of Future Coexistence, 16.
33 Morton, 16.
34 Morton, 17.
35 Young, "On Correlationism and the Philosophy of (Human) Access," 50. Emphasis added.
36 Collins, "We Cannot Live by Scepticism Alone."
37 Collins.
38 Collins, Evans, and Gorman, "Trading Zones and Interactional Expertise"; Galison, "Trading with the Enemy."
39 Berrizbeitia, "Scales of Undecidability."
40 Berrizbeitia, 124.
41 Berrizbeitia, 125.

Bibliography

Bennett, Jane. *Vibrant Matter: A Political Ecology of Things*. Durham, NC: Duke University Press, 2010.

Berrizbeitia, Anita. "Scales of Undecidability." In *Downsview Park Toronto*, edited by Julia Czerniak, 117–25. CASE Series (Prestel Verlag). Munich; London: Prestel, 2001.

Cantrell, Bradley, and Justine Holzman. *Responsive Landscapes: Strategies for Responsive Technologies in Landscape Architecture*. New York: Routledge, 2015.

Collins, Harry. "We Cannot Live by Scepticism Alone." Comments and Opinion. *Nature* (March 4, 2009). https://doi.org/10.1038/458030a.

Collins, Harry, Robert Evans, and Michael E Gorman. "Trading Zones and Interactional Expertise." In *Trading Zones and Interactional Expertise: Creating New Kinds of Collaboration*, edited by Michael E. Gorman. Inside Technology. Cambridge, MA: MIT Press, 2010.

Cross, Nigel. "Designerly Ways of Knowing: Design Discipline Versus Design Science." *Design Issues* 17, no. 3 (July 2001): 49–55. https://doi.org/10.1162/074793601750357196.

Davis, Brian. "Landscapes and Instruments." *Landscape Journal* 32, no. 2 (2013): 293–308.

Galison, Peter. "Trading with the Enemy." In *Trading Zones and Interactional Expertise: Creating New Kinds of Collaboration*, 25–52, 2010.

Graeber, David, and David Wengrow. *The Dawn of Everything: A New History of Humanity*, 1st ed. New York: Farrar, Straus and Giroux, 2021.

Halprin, Lawrence. *The RSVP Cycles: Creative Processes in the Human Environment*, 1st ed. New York: George Braziller, 1970.

Hui, Yuk. "Machine and Ecology." *Angelaki* 25, no. 4 (July 3, 2020): 54–66. https://doi.org/10.1080/0969725X.2020.1790835.

Klosterwill, Kevan. "The Shifting Position of Animals in Landscape Theory." *Landscape Journal* 38, no. 1–2 (January 1, 2019): 129–46. https://doi.org/10.3368/lj.38.1-2.129.

Luhmann, Niklas. "How Can the Mind Participate in Communication?" In *Theories of Distinction*, translated by William Rasch, 169–86. Redescribing the Descriptions of Modernity, Cultural Memory in the Present. Stanford, CA: Stanford University Press, 2002.

Lystra, Margot. "McHarg's Entropy, Halprin's Chance: Representations of Cybernetic Change in 1960s Landscape Architecture." *Studies in the History of Gardens & Designed Landscapes* 34, no. 1 (January 2, 2014): 71–84. https://doi.org/10.1080/14601176.2013.850313.

Maturana, H. R., and F. J. Varela. *Autopoiesis and Cognition: The Realization of the Living*. Boston Studies in the Philosophy and History of Science. Cham: Springer Netherlands, 1980. https://doi.org/10.1007/978-94-009-8947-4.

Meyer, Elizabeth K. "Situating Modern Landscape Architecture (1992)." In *Theory in Landscape Architecture: A Reader*, edited by Simon R. Swaffield, 21–31. Penn Studies in Landscape Architecture. Philadelphia: University of Pennsylvania Press, 2002.

———. "The Expanded Field of Landscape Architecture." In *Ecological Design and Planning*, edited by George F. Thompson and Frederick R. Steiner, 45–79. New York: John Wiley & Sons, Inc., 1997.

———. "The Post–Earth Day Conundrum: Translating Environmental Values into Landscape Design." In *Environmentalism in Landscape Architecture*, edited by Michel Conan, Vol. 22, 187–244. Washington, DC: Dumbarton Oaks Research Library and Collection, 2000.

Morton, Timothy. *Dark Ecology: For a Logic of Future Coexistence*. The Wellek Library Lectures in Critical Theory Wellek Library Lecture Series at the University of California, Irvine. New York: Columbia University Press, 2016.

Reed, Chris, and N.M.E. Lister. *Projective Ecologies*. Actar D, 2014.

Spirn, Anne Whiston. "Ian McHarg, Landscape Architecture, and Environmentalism: Ideas and Methods in Context." In *Environmentalism in Landscape Architecture*, edited by Michel Conan, 97–114. Washington, DC: Dumbarton Oaks, 2000.

Steiner, Frederick R. "The Ghost of Ian McHarg." *Log*, no. 13/14 (2008): 147–51.

Young, Niki. "On Correlationism and the Philosophy of (Human) Access: Meillassoux and Harman." *Open Philosophy* 3, no. 1 (January 1, 2020): 42–52. https://doi.org/10.1515/opphil-2020-0003.

CONCLUSION

Design and Cybernetic Environment

This book argues for a paradigm of cybernetic environments as an alternative to cybernetic systems. The two can be contrasted by using the terms in the list below.

Cybernetic System	*Cybernetic Environment*
Sensing-modeling-actuating	Coding-choreographing-attuning
Anthropocentrism	Posthumanism
Extension of human agency	Co-productive/distributive intelligence
Co-dependency	Interdependency
Controlled stability	Cultivated wildness
Failure	Machine perspective
Uncertainty	Possibility
Technological monoculture	Technodiversity
Wild nature	Wildness in machines
Transformation	Co-production
Scales of undecidability	Scales of speculative ecology

Cybernetic Systems: Stability by Machines and Wildness by Nature

The cybernetic systems paradigm is rooted in the humanistic principles inherited from the Enlightenment and celebrate human agency and the ability to break free from established beliefs, laying the groundwork for the political revolutions that followed in America, France, and other nations. However, this focus on human exceptionalism has also led to an unwavering faith in

DOI: 10.4324/9781003320852-19

human control over other entities, whether biological or technological. The concept of agency is limited to humans, whereas intelligent machines are seen as mere tools that can extend, amplify, and replicate human physical and mental capacities. As a result, our relationship with intelligent machines is one of co-dependence, as they are viewed as extensions of human agency. Unfortunately, nonhuman agency is completely outside our conceptualization. This undertaking extends the colonial transformation formula investigated in Part I; humans use technology to transform nature into habitable landscapes. Cybernetics only adds another preeminent tool to this outdated framework as the ultimate generalization of our relationship with other nonhuman beings.

When we consider the impact of human agency on the environment, we tend to view it in terms of a spectrum between control and wildness. The term "wildness" here refers to unpredictability and uncertainty beyond human understanding. We often refer to the nonhuman realm as "wild nature" because we continually encounter new patterns and relationships outside our comprehension. At one end of the control-wildness spectrum is a state of complete human control where potential uncertainties are regulated to maintain a predictable environment, such as a city. At the opposite end lies ultimate wildness, such as in the Amazon rainforest and the Great Barrier Reef in the eyes of an urban dweller, where human agency appears irrelevant. In the center lie different degrees of perceived human agency. In this conceptualization, wildness and control are mutually exclusive and negatively correlated; control would vanquish uncertainty, thus reducing unpredictability and wildness.

Many contemporary environmental ideas explored in this book can be placed on the control-wildness spectrum. The Anthropocene, a popular concept that underpins many environmental narratives, highlights the scale of technological systems in influencing the environment. Technology is viewed as a means of exercising control over it. For over 70 years, cybernetics has been reduced to a techno-scientific endeavor; the sensing-modeling-actuating feedback loop is a means for controlled stability. Certain environmental narratives advocate scaling back the techno-scientific imagination, as seen in concepts like "planetary boundaries" and "half-earth," and allowing nature to function independently. Alternatively, the Anthropocene can be embraced by upscaling our techno-scientific imagination to construct a "nature" that supports human survival, as advocated by ecomodernists. This control-wildness spectrum also manifests in smart cities, which regard technology as a layer of "digital infrastructure" regulating and optimizing urban systems; by making cities "smarter," wild nature can be saved.

Cybernetic Environment: Cultivated Wildness

The cybernetic environment paradigm is not "environmental cybernetics." The latter still falls in the paradigm of cybernetic systems. It black-boxes environmental phenomena into a cybernetic system while simultaneously constructing an external environment. This view is rooted in a system-environment dualism—that which is outside the black box is reduced to a homogeneous backdrop accessible only by flows of input and output. In turn, the homogeneous environment defines how we individualize an agent, whose agency and intelligence become localized capacities restricted by the imagined boundary. This conceptualization is incompatible with contemporary posthumanism's concerns for distributive agency.

By re-conceptualizing cybernetics, I propose a cybernetic environment paradigm, which asserts that a system's outside environment is a heterogeneous meshwork of objects, assemblages, and mental processes. From this vantage, defining the environment as a "space" is problematic by rendering it into an inactive backdrop, as if nothing occurs. A system cannot interact with a void but only with other assemblages. Therefore, if all objects, assemblages, and mental processes were removed, no "environment" would exist. A cybernetic environment thus speaks to the totality of the meshwork of different assemblages and objects, redefining what an "environment" entails—to be lively, not empty.

A cybernetic environment defies inside-outside and system-environment dualisms. It seeks to overcome the urge in system thinking to black-box more objects into a cybernetic system while producing yet another exterior environment. This type of system thinking is still anthropocentric, no matter how many nonhuman objects we summon to mold the human world. We must overcome the tendency to view a collection of processes as a cybernetic system that motivates strategies for communication and control. Perhaps this tendency is useful in designing and building powerful machines, but systems thinking leads us only so far. This book suggests alternative ways to frame the environment as a *cybernetic environment* rather than a *cybernetic system.*

Cybernetic thinking has always dealt with cybernetic systems, not cybernetic environments. From this vantage, many concepts, such as adaptive management, responsive landscapes, cyborg ecologies, and smart cities, should be categorized as "environmental cybernetics" operating within the old paradigm of cybernetic systems. These approaches apply cybernetic principles to optimize and control environmental processes by first imagining them as systems. A smart city conceptualizes urban spaces as systems with computable variables and produces an exterior as an environment, which then presents systems with a new set of uncertainties to be reduced with other cybernetic systems. We are confined in this cycle of using one system to replace the other—a continuation of progress narratives that extend imagined human agency through a systemic construction.

Though many ideas, such as responsive landscapes, avoid terms such as *control* and embrace emergence and open-endedness to emphasize a "lack of control," they cannot do so without first articulating the environment as an "open system" and regarding feedback mechanisms as a means to achieve emergent behaviors. Over the past three decades, the discipline of landscape architecture has developed a process-based design framework, articulating landscapes as open systems that evolve, unfold, and thus become open-ended. This conceptualization undergirds today's landscape theory and practice. However, process-based frameworks do not overcome the cybernetic system paradigm; they perceive landscapes as evolving systems communicating with their outside environments. This is why contemporary process-based landscape design struggles with the notion of open-endedness. The irony here is that feedback mechanisms become yet another control strategy, chosen to intentionally produce run-away behaviors in an emergence framework. Designers still control the system toward an open-ended direction, and system emergence is instrumentalized as a means to another end.

In their book *Responsive Landscapes*, landscape architects Cantrell and Holzman asked, "[a]t what point are goals, scaffolds, and protocols actually open-ended?"[1] This question cannot be resolved within the paradigm of cybernetic system, because as long as we conceive of a system, we construct its environment. Whatever lies outside this "landscape system" and undermines its emergent behaviors becomes a new threat that challenges designers' urge to control. Only when we shift our attention from systems may we begin to develop alternative framings. Contemporary landscape theory and practice, rooted in process-based strategies, have sought to explore an alternative framing. However, this exploration will not bear fruit until one takes on a deeper ontological and epistemological reconfiguration that surpasses the cybernetic system as the only way to interpret cybernetics itself. On this level, this book prizes other ways of reasoning rooted in contemporary posthumanist concerns on nonhuman agency, intelligence, and inaccessible "surplus" in objects. The environment is a cybernetic environment—a meshwork of different assemblages and objects interacting with each other through feedback mechanisms and recursive processes. It results from an amalgamation of different goals, mental processes, and objective functions distributed across the meshwork. This is not an argument, but an ontological claim which one must subscribe to and then utilize to generate moral and ethical considerations.

How, then, might designers participate in a cybernetic environment comprised of non-communicative and uncontrollable objects and assemblages? What does design mean in a cybernetic environment? Many landscape architects and theorists, such as Brian Davis and Anita Berrizbeitia, endeavor to answer these questions by articulating a space between a designer's intention and the reality co-produced by more than human beings.[2] The environment becomes a place with different levels of human intent.

However, this reasoning pushes the process-based landscape framework too little. One may still conceptualize what is influenced by the designer's intentional decisions as a cybernetic system, unfolding and evolving under the designer's control, while the uncontrollable poses challenges to designers' goals, even if the aim is to promote open-endedness and emergent behaviors. Thus, the liminal space between intention and reality leaves a great deal of room in which designers may expound on their agency and hubris. This is why present-day landscape theory and practice still prioritize knowing, particularly human knowing, as the foundation of design. This urge to *know* motivates designers to define new systems. Thus, the most profound irony in today's nonhuman turn within landscape architecture is that, even though designers wish to embrace a posthumanist framework that de-centers designers from the source of agency, the cybernetic system paradigm still propels designers to summon more nonhuman objects to participate in our stories of controlled emergence.

The predicament in contemporary landscape theory and practice is to view intentionality and unintentionality as mutually exclusive, with shades of gray in between. Instead, I have offered in this book a different framing. When engaging with a cybernetic environment, we must not understand intentionality by percentage. Because the environment is a collection of non-communicative and uncontrollable objects that occasionally synchronize and attune to each other, we cannot hope that designers' intentions can be actuated at all. Objects constantly reduce each other within their own system operations, and respond in their own ways. Thus, designers must recognize that any intentional decisions may be modified, amplified, dampened, or canceled by other actors. Designers' intentions constantly create unintended consequences—*cultivated wildness* is the metaphor which I propose to describe this new condition in which design should be framed. In other words, a cybernetic environment flips what is normal and constant; the so-called intentional outcomes—those situations in which we believe we are in control—now become flickering moments when things accidentally synchronize and align. Uncontrollability and non-communication become normality. Consequently, design becomes a constant effort to cultivate attunements between human and nonhuman objects and assemblages, hoping things might accidentally align.

In contrast with a cybernetic system or "environmental cybernetics" that relies on a control-wildness spectrum, the cases we confronted throughout Part III offer an alternative framing for cybernetics through a coding-choreographing-attuning feedback loop to replace the sensing-modeling-actuating one. This new framing emphasizes three different aspects concerning environmental design.

1 Sensing networks do not innocently listen to the environment; instead, designers must code the environment into information that makes a difference.

Using coding to replace sensing emphasizes an adaptive epistemology, which differs from the constructivist view that all knowledge is socially and culturally constructed. *Coding* highlights co-production between humans and the nonhuman realm. The way in which we represent the environment cannot be separated from the way we choose to live in it.[3] This means that sensors are no longer passive listeners, but instruments that aid us in living within the assemblages of actors. Sensing is a practice that involves encoding phenomena with instruments, and, as such, designers must code the environment to address urgent issues, such as climate change and social inequality, and their compounding effects on both the human and nonhuman communities. In light of our increasing environmental uncertainty, sensing practices must go beyond data collection. Designers working with cybernetic technologies must leverage their expertise to challenge conventional environmental sensing practices and promote technodiversity to unlock the untapped potential of sensing networks.

2 Designers should move from constructing models of the environment, and instead work with intelligent agents in a non-model-centric paradigm to explore the possibilities offered by these nonhuman agents.

By using the term "choreographing" instead of "modeling," designers can establish a different relationship with their models, especially in light of the advancements in machine-learning techniques for environmental practices. Modeling constructs a mathematical representation for the system being modeled, which highlights the gap between the model and the actual system. The model-centric paradigm, which is the basis of modern control theory, emphasizes the importance of model accuracy for the success of control strategies. This paradigm propels designers to focus on improving models to reduce the gap between the model and the system while overlooking the primary purpose of creating the model. Within this framework, designers are viewed as individuals who rely on digital models and algorithms to enhance their intelligence and expand their perceived influence through mechanical repetition.

In contrast, "choreographing" emphasizes a relationship built on trust. As discussed in Chapter 12, recent advancements in modeling techniques have led to a non-model-centric paradigm, where designers and models co-produce and co-evolve. Within this framing, machines are employed as collaborators rather than mere tools for extending perceived human agency. Unexpected behaviors that arise from machine learning become valuable lessons that offer a unique perspective to challenge and inspire our all-too-human ways of thinking and doing. The relationship between designers and algorithms transcends the traditional creator-created dynamic and is better described as that of choreographers and dancers. This perspective renders the notions of human intelligence and machine intelligence irrelevant, as what emerges is a collaborative intelligence resulting from the co-production process.

3 Repeated actuation does not result in controlled stability, but instead leads to a cultivated wildness in a complex meshwork of goals, frameworks, and objective functions which we refer to as "the environment."

By replacing the term "actuating" with "attuning," designers can embrace a developmental perspective on the relationship between intelligent machines and human and nonhuman actors. In modern control theory, actuators implement control policies to achieve a desired state, resulting in controlled stability. However, in scenarios such as Sougwen Chung's drawing operations, repeated actuation can lead to unexpected behaviors in unfamiliar territory, creating a cultivated wildness. Here, control and wildness are not oppositional and mutually exclusive. At the center of this framing is the conviction that the environment is a meshwork of mental processes, frameworks, goals, and objective functions, resulting from co-production between human and nonhuman actors. The outcome is alien to all, brimming with wild situations. Designers can collaborate with intelligent machines to explore this untamed territory, where cultivated wildness represents a reserve of possibilities and strategies for an unforeseen future.

Prototyping, Theorizing, and Transdisciplinarity

This book foreshadows a design research paradigm characterized by prototyping and theorizing as the two venues for a transdisciplinary practice in the cybernetic environment.

The cases in Part III highlight the importance of prototyping as a preeminent tool for generating new framings of technologies and techniques. It is worth noting that prototyping research differs from the more common idea of product prototyping in industrial and software design. Here, prototyping research involves denaturalizing a machine or a technique by situating it in different scenarios to generate new framings. For instance, Chung's drawing operations produced prototypes of drawing machines that may not be talented painters, but the interaction between the artist and the robot informs a framing of robots as collaborators rather than mere tools. Other examples include the hydromorphology table and the gardening robot used in landscape research. These prototypes provide a glimpse into an ecosystem of computerized ecology where the relationships between humans, plants, and other nonhuman species are deeply mediated and, at the same time, enabled by sensors, controllers, computer hardware, layers of computer code, and online servers.

Although these prototypes cannot be applied directly in practice, they are crucial in supporting transferable and generative contributions to knowledge, rooted in ad hoc yet innovative methodologies.[4] As physical objects, they encapsulate unique techniques that contribute to technodiversity, especially in a technological "landscape" dominated by "monoculture." The value of prototyping research lies in its capacity to explore technodiversity and generate inexhaustible meaning in various contexts. Ideally, the interpretational flexibility of prototypes could keep generating knowledge beyond their original scope.

In recent years, an increasing trend among architecture and design schools has been the establishment of design laboratories equipped with advanced machinery, fostering a "lab culture" within design disciplines. These labs have produced a multitude of prototypes, prompting important questions about how they should be documented, discussed, and theorized within appropriate frameworks. Moreover, how can prototypes be analyzed and reflected upon in a way that generates knowledge beyond the design labs and is relevant to modern disciplines? Additionally, what distinguishes the expertise of these design labs from that of scientific laboratories? These questions call for further investigation and reflection, as they are essential to understanding the significance and impact of design labs on contemporary design practices.

One promising approach to addressing these questions is to embrace non-dualistic thinking and move beyond the traditional dichotomy of theory and practice. Theorizing, in this view, is itself a practice that connects ideas across different domains. Landscape design, for instance, is a practice that embodies theories about the world in built forms and material assemblies. As landscape theorist Elizabeth Meyer argues, landscape theory's role is bridging, mediating, and reconciling between thought and action, between a particular project and a general principle, or between a general principle and a specific site.[5]

The four types of coding practices presented in Chapter 11 offer examples of how to extract general principles from diverse environmental sensing practices, even if they are not specifically related to landscape design. By theorizing about these practices and using terms such as cloning, assembling, rerouting, and physicalization, their conceptual strategies can be preserved and highlighted, and then "ported" to other fields. Porting is a software-engineering term that refers to the process of adapting software from one computing environment to another, while maintaining its functionality. Similarly, theorizing is a practice that involves porting ideas from one realm to another, where the interpretative context may differ, but the conceptual functions remain intact. In this sense, theorizing is inherently transdisciplinary.

Transdisciplinarity is essential in the cybernetic environment, as cybernetic thinking underpins most modern disciplines. Prototyping and theorizing are two powerful tools with which to navigate and explore diversity of thought. In this exploration, cultivated wildness offers an alternative version of cybernetic thinking based on a speculative ontology that recognizes beings' non-communicative and uncontrollable natures. All modern machines are cybernetic machines, but as objects, they are withdrawn and inaccessible, so we can only speculate about the types of dynamic living relations we can form with them. Therefore, we must adopt a speculative perspective and reframe machines in different conceptual frameworks to explore their wildness. Only in this way may we use machines as media to form different living and dynamic relations with other beings.

If modern technology, according to Heidegger, is *enframing,* which posits everything as a standing reserve or resources to be exploited, then "wildness in machines" takes a positive turn by framing machines within a speculative ontology. If we begin with the assertion that things, including machines, are withdrawn and can only be speculated upon, then nothing can be completely accessed and thus fully exploited. Exploring wildness in machines is to explore possible relations with beings around us. Only in this way can the cybernetic environment become a reserve of possibility, filled with speculative ecologies between withdrawn objects.

Notes

1 Cantrell and Holzman, *Responsive Landscapes,* 254.
2 Davis, "Landscapes and Instruments"; Berrizbeitia, "Scales of Undecidability."
3 Jasanoff, "The Idiom of Co-Production."
4 Arrigoni, "Epistemologies of Prototyping."
5 Meyer, "Situating Modern Landscape Architecture (1992)."

Bibliography

Arrigoni, Gabriella. "Epistemologies of Prototyping: Knowing in Artistic Research." *Digital Creativity* 27, no. 2 (April 2, 2016): 99–112. https://doi.org/10.1080/14626268.2016.1188119.

Berrizbeitia, Anita. "Scales of Undecidability." In *Downsview Park Toronto*, edited by Julia Czerniak, 117–25. CASE Series (Prestel Verlag). Munich; London: Prestel, 2001.

Cantrell, Bradley, and Justine Holzman. *Responsive Landscapes: Strategies for Responsive Technologies in Landscape Architecture*. New York: Routledge, 2015.

Davis, Brian. "Landscapes and Instruments." *Landscape Journal* 32, no. 2 (2013): 293–308.

Jasanoff, Sheila. "The Idiom of Co-Production." In *States of Knowledge : The Co-Production of Science and the Social Order*, edited by Jasanoff Sheila, 1–12, 2004.

Meyer, Elizabeth K. "Situating Modern Landscape Architecture (1992)." In *Theory in Landscape Architecture: A Reader*, edited by Simon R. Swaffield, 21–31. Penn Studies in Landscape Architecture. Philadelphia: University of Pennsylvania Press, 2002.

INDEX

Note: *Italic* page numbers refer to figures.

FSC
www.fsc.org
MIX
Papier | Fördert gute Waldnutzung
FSC® C083411

Kolibri 360 GmbH
Ferdinand-Jühlke-Straße 7
99095 Erfurt, Deutschland
produktsicherheit@kolibri360.de